OPTIMAL DISTANCE

OPTIMAL DISTANCE

A Divided Life

PART TWO

Publisher: Camperdown Elm Publishing, LLC, P.O. Box 39, Boulder, Co. 80302

First Edition: September 2017

The Bibliography and Biographical Index for *OPTIMAL DISTANCE, A Divided Life, Part One and Part Two*, can be found at the author's website, www.optimaldistance.com.

The names and identifying characteristics of a few individuals have been changed, whether so noted in the text or not.

ISBN 978-0-9987690-2-8

Library of Congress Control Number: 2017947063

Designed by T. Keith Harley

Pre-publication preparation by Scott S. Miller

Printed in the United States of America

TOPAZ MOUNTAIN – UTAH

circa 1890

Contents

To the Animals Most Like Us

Bear mother nursing her twin cubs in Yellowstone.

All my life, animal companions have provided me with comfort and guidance, frequently touching the deepest parts of my soul. I do not believe that I would have survived long enough to excavate my life and document my personal history without their companionship. Fifty years after Commander Margaret insisted that my father shoot our two springer spaniels, Army and Nurse, they returned to my side in my dreams and gently nudged me toward the truth. The devotion of animals, their GPS-like intelligence, and their ability to trust other species, even humans capable of brutal betrayal, are why cross species relationships remain my greatest source of hope for our world. When animals shrink the distance, accepting the differences between themselves and other species, they are showing us how to achieve *optimal distance* and mend our broken world.

– PREFACE TO PART TWO –

Second Chances and the Excavation of a Life

From birth to thirty-nine, my mother's mind had ascendancy over mine. Her death on the morning of my fortieth birthday was more like an amputation that left behind a phantom limb still sending alarm signals to my brain. My rational self knew she was dead; I had stood by her body as the coroner signed her death certificate. Even more to the point, at her bequest I had attempted to repair her mortuary make-up, working on her embalmed face as her body lay inside the casket, an awful task. My mother's casket contributed in part to her phantom-like presence. Selected for both economy (my father's) and for irony (mine), it was a simple wooden box, hand-pegged to meet the "dust unto dust" requirement of Jewish burial customs (Genesis 3:19). But without any metal, not even a single nail, I realized the casket was an inadequate container for my mother's fierce power. Margaret Audrey Beck Lieberman was dead, but her shadow felt indestructible, immune to interment.

A hard rain fell as her casket was slowly lowered into the plot she had secretly purchased. Desperate to ensure my mother would remain at an *optimal distance* from me for eternity, I asked the four gravediggers to stand aside. I was quickly up to my ankles in mud as I struggled to fill her grave with six feet of Colorado clay. Even as those heavy shovelfuls fell onto her casket, I imagined that she was breaking out of the flimsy wooden container, after which she would use her hands and nails to scrape away the clay. Perhaps she had found the secret to resurrection in her obsessive study of the Dead Sea Scrolls. Alternatively, since my mother often claimed God was talking to her, there was an extremely remote possibility that He was on his way, bringing her the perfect auger.

The death of my mother was not what led me to take a second chance on motherhood. My husband and I had already made that decision, but her absence made me more eager to start my life over. So, twenty years after I gave birth to Olivia, and a year after my mother's death, I gave birth to a son. For the first three years of his life, I felt as if living was a secret that had been kept from me. I couldn't tell whether it was his presence or my mother's absence, but I had never felt happier, as if I was swimming in oxytocin. Maybe I was because I was still nursing when I was diagnosed with breast cancer.

My son was six when the breast cancer metastasized to my liver. Trembling as I sat across from my oncologist, I asked how much longer I could expect to survive.

"About six months, if you are lucky and have a good response to treatment," Dr. T said, writing notes, avoiding eye contact.

Weak with fear, I accepted Dr. T's prognosis, packed my bag, and began the anxious wait for death's knock on my door. Death was such an important trip! I crossed my fingers, hoping my escort would be kind. I was shocked, rather than relieved, when death kept me waiting. Before my diagnostic death sentence, my focus had been on looking forward in time to find hope. But with death in close proximity, I felt an overwhelming urge to go back in time in search of meaning.

Examining my past was not easy. I often felt like a naked rescue worker digging through mounds of rubble. There were many shards in the rubble—some were fragile and quite beautiful, others were dangerously sharp, and many felt too filthy to touch. The work was exhausting, but the search for meaning required the careful examination of whatever was unearthed. There were many times when I became all but buried in the rubble of my life.

As I proceeded with my archaeological dig, the cells in my body responsible for its defense slowly began to wake up. A previously protective amnesia began to dissipate in dreams. Finally the murderous beast of my early childhood emerged from the fog of my unconscious, demanding that I look it in the eye. I resisted until, in a shivering coincidence, Mother Nature intervened. Forced to keep digging deeper for long buried secrets, I was only able to believe what my unconscious was telling me after hearing the testimony of three witnesses.

Now I believe that seeing the true shape of my life was what contributed to a spontaneous radical remission; a commotion occurred at a cellular level that activated my battered immune system. While there is no test to measure these types of mysterious changes, I believe whatever happened was linked to my conscious mind finally being able to face the terror I experienced on The Day the Bear Went to Topaz (see *Part One*). While my remission didn't last forever, it nevertheless contributed to my long term survival, giving me time to explore and be converted to the new science of epigenetics.

While there is a growing body of literature describing cases of radical and spontaneous remissions, as well as the ways in which many patients have managed to turn metastatic cancer into a chronic condition, the challenges of an unexpected lengthy survival are largely untouched. I have given up all hope more than once. Four months after listening to Dr. T's prognosis in 1989, I became a hospice patient having "failed chemotherapy" and fallen apart. Three months later, I entered a clinical trial which made it necessary for me to check out of hospice. Then I began a rigorous metabolic treatment, had a complete hysterectomy, and entered yet another clinical trial, etc. I have had more slash, burn, and poison treatments, as well as more carrot juice, dialysis, and surgery than the population of a small town or some cities in Africa--a strange embarrassment of riches.

Finding *optimal distance* from death is a lifelong challenge for everyone. Most of us prefer to skip any previews of our inevitable mortality. My experience has taught me that when the body begins to fail for whatever reason, it is best to prepare for both outcomes—a longer-than-expected period of survival or a sharp decline. The mistake I made was to over-prepare for imminent death and under-prepare for an extraordinarily long period of survival.

Because my life has been filled with an abundance of coincidences, my death wish is for scientists to discover more about the cellular and chemical phenomena that store unconscious memories and deliver dreams filled with guidance. I will die with both an abiding respect for the wisdom of our unconscious minds and filled with hope for the future of epigenetics. I am deeply grateful for those forces in the universe that extended my life beyond scientific explanation, granting me time to undertake my search for meaning, to describe my discoveries, and to make peace with all that remains unknown.

* * *

– BOOK III –

SECOND CHANCES

A problem is something which I meet, which I find complete before me, but which I can therefore lay siege to and reduce. A mystery is something in which I myself am involved.

Gabriel Marcel, *Being and Having* *

* Marcel, Gabriel. *Being and Having*. London: Dacre, 1949, p. 100.

-1-

Crossing Over to a New Life

Olivia left for college a year before my mother died. Her absence made my body feel as if something essential had been scraped out. I missed her lively presence at the dinner table and the sound of her laughter as she spoke with friends on the telephone. I even missed her rustling through my closet or interrupting my meditative bedtime bath with a question or concern that couldn't wait.

Within a month of Olivia's departure, I began losing my hair and having repetitive dreams of Bob holding an infant. Dreaming of having a child with Bob, wanting to be a parent with a man I loved as I had no other, to have him as a full and equal partner, was a complete surprise. My unconscious desire was undoubtedly triggered by Olivia's absence, as well as by the persistent tick of my hormonal clock. What is the biological source of such impulses at forty? Was this a female mid-life crisis? Was the desire coming from my lonely maternal heart or my waning hormones—or something else?

Three years before I married Bob, I had a tubal ligation, making a final choice not to have another child. When I learned a new microsurgery technique might successfully reverse that decade-old decision, I had cautiously raised the possibility with Bob. He was adamantly against having another child. Not only did he not want to return to the intense dependency of those early years, he felt we did not have the necessary financial resources.

With great certainty, Bob told me, "This is not something you want to do. I know you, Joan. Having a baby now is not right for you!"

I then had several clients who were women in leadership positions in government organizations, law schools, and large professional practices. Now that there were just two of us at the dinner table, and no school day or homework discussions, we spoke more frequently about our separate worlds of work, including about the unconscious organizational misogyny my female clients were facing—most of which I had experienced in my own work and consulting roles.

Soon after Bob's expressed opposition to having a baby, he spent an evening with our friends Monroe and Aimée Price in Los Angeles. Monroe had just been offered the deanship of Cardoza Law School in New York City. He wanted to accept the position, but Aimée was reluctant to leave her native Southern California. Monroe insisted that the move to New York would be good for Aimée's career as an art historian. Bob caught a glimpse of himself in the mirror of their marital dialogue. He realized in arguing against trying to conceive another child, he was using the same "I know what is best for you" argument as Monroe was using with Aimée. Bob returned to Colorado and gave me his permission.

On June 1, two weeks before my mother's death, I saw Dr. Manfred Oliphant in Denver and scheduled the microsurgery for July 2. After my mother's death, as I packed my bag for the hospital stay, I realized I had not told her that we were going to try to have another child. Then I wondered if she knew anyway.

In mid-July, after the surgery, Bob and I invited my father to live with us while he adjusted to his new life as a widower. The three-bedroom apartment on Cedar Avenue suddenly seemed too spacious and too spooky. Jordan was also spending the summer with us. His twelfth birthday had been celebrated the night before the surgery. Watching Jordan blow out the candles on his cake, I realized that since the winter holidays, he had grown almost foot taller—his blond hair and long eyelashes were the envy of every female in his orbit.

Jordan on his twelfth birthday, Boulder, July 1, 1982

Both my father and Jordan seemed at loose ends, each transitioning to new stages of life as widower and teenager. When Aunt Mary called to invite me to an August reunion in Utah of the Beck family, I persuaded the two restless males to accompany me to the reunion via Yellowstone National Park. Jordan, having heard some of our family bear tales, was anxious to see some Yellowstone bears for himself.

In early August, Frank, Jordan, and I packed our things into my father's 1972 Toyota station wagon and drove north. We started Jordan's determined hunt for bears by trekking around Jenny Lake in Teton National Park.

Hearing a long low growl, Jordan asked excitedly, "What was that? Did you hear that?"

"Yeah, I'm sure it was a bear," my stone-faced father replied.

There was no bear; Frank's bowels had been stimulated by the hike. When Jordan caught on to his step-grandfather's joke, there was no way to stop further conversational decline. Jordan was of an age when bodily functions rise to the top of the male agenda. For my father and me, I now wonder if it wasn't an unconscious attempt to heal our wounded memories of my mother's imaginary poisonous snakes.

On our last day in Yellowstone, we rose before dawn to fish the Madison River where we met another fisherman who had seen a bear by Heart Lake the previous day. We didn't have time to go back to Heart Lake, but we modified our plans. With fishing poles and day packs, we headed north into the back country on the Gneiss Creek Trail at the west end of Madison Canyon. Jordan more than met the challenge, even though it was the longest distance he had ever hiked. Our trio made it all the way to the Montana border. We fished Campanula Creek, gently releasing our collective catch of two tiny trout. We saw considerable bear scat, but no bears. That night we crawled into a West Yellowstone motel and slept as if hibernating.

In the morning, I took a photograph of Jordan standing next to a large grizzly bear in a West Yellowstone gift shop. Jordan made us sign a secret pact to tell everyone at home that the grizzly was alive, instead of an example of expert taxidermy. When Jordan called Bob that night, I overheard him previewing his tall tale.

"Dad," Jordan said, "we finally saw a bear. A really big one! I've got pictures too!"

After crossing the Idaho-Utah border, we stopped for lunch at the edge of Bear Lake, where Frank taught Jordan how to skip stones—the same shore where he had taught me thirty years earlier. Then we made our way down Logan Canyon on Hwy 89 to my childhood hometown. Before checking into a motel, we drove past our former homes. All the structures were still standing, but sadly the white bricks of my mother's Dream House had been painted sage green and partially covered with wood siding. There was no evidence of my father's handcrafted burgundy shutters.

Afton Evans was sitting on her porch when we parked in front of our old basement apartment. Even smaller at age seventy-seven, Afton was still wearing her hairnet. Although we had been absent from her life for twenty-five years, she recognized us immediately and graciously invited us inside for glasses of Welch's grape juice.

As we stepped into the living room, I couldn't believe my eyes! There was Ben's birthday couch, still in perfect pale satin brocade condition, finally freed from its stiff protective plastic cover. Ben had died in 1976. Remembering my first independent prayer on his behalf, I smothered my desire to ask Afton whether she had ever allowed Ben to take a nap on his birthday couch without the plastic cover.

Afton declined our invitation to join us for dinner at the Blue Bird Café, but insisting we visit Marlene, who was married and living in a Salt Lake suburb with her husband and three children.

After dinner, my father and Jordan were both ready to go to bed in their shared room. Unable to sleep in my solo room, I took a long walk up to the campus of Utah State University and beyond. Circling through Hillcrest, past the former homes of Scott Olsen and Elizabeth Budge, now miniature replicas of the edifices stored in my memory.

Walking back to the motel, I stopped at the artfully-lit Logan Temple, the site of my pseudo-solo baptism. Remembering my secret shame during that era, I realized that early religious indoctrination is impossible to completely eradicate. We can lose our childhood faith, but indoctrinated fear seems to follow forever.

Returning to Main Street, I found a bench on the grounds of the Logan Tabernacle and took a seat until the robins began singing at dawn. Only after the sun came up did I realize Wickel's Men's Store was gone.

After an early breakfast, Frank drove through Brigham Canyon toward Salt Lake City. We stopped for lunch at the Maddox Ranch House restaurant to touch another memory receptor with a taste of the restaurant's phenomenal fried chicken. Jordan and I savored every bite, while my father stoically ate a special order of steamed trout. After lunch we searched Ogden for any evidence of my father's childhood, but it had been obliterated.

The sun was sinking into the Great Salt Lake when we checked into the still grand Hotel Utah. In the dining room where Frank had proposed to Margaret, he wept through dessert after I asked him what he recalled about those magical moments.

Dinner done, the three of us walked around Temple Square, smiling at the circulating missionaries while politely refusing their proffered literature. Long past Jordan's bedtime, we met Olivia at the airport. She had flown over the Rocky Mountains to join us for the Beck reunion.

In the morning, our foursome drove to the home of Marlene Evans. She looked remarkably the same, and her daughters were replicas of my childhood friend.

I was stunned when Marlene told me, "You know, Carol, I still have the Raggedy Ann doll your mother made for me!"

She sent one of her daughters to retrieve it. Holding my mother-made doll, I began weeping. I wanted to ask Marlene to let me have it after all these years, but pushed back against my regressive longing. Our reunion ended with ritual picture-taking, including a self-timed shot of the full group using the roof of the Toyota station wagon as a tripod.

Many miles later, as we were nearing the site of the Beck family reunion at a park in Bountiful, I realized I had left the camera and my address book on the roof of the Toyota. Frank pulled to the side of Hwy 89, but there was nothing except dust and dead insects on top of the station wagon. We placed a lost and found notice with a reward in the *Salt Lake Tribune*, without response. Sadly, the images of Jordan's Yellowstone bear and of Marlene and her progeny were stored only in our vulnerable minds.

The Beck's family reunion was disappointing. Most of the cousins my age seemed to be attending under duress and made little attempt to reconnect. My excommunication from the Mormon Church had more than doubled the distance between us.

My mother's oldest sister, my Aunt Vermilla, sat at a table covered with displays of Nanie's needlework and one of the welding helmets Pepa invented, next to a large flannel board covered with family photographs. After her fiftieth birthday, Aunt Vermilla had gone back to school for a degree in nursing.

Taking a seat next to her, after a few minutes of small talk, I asked, "Do you have any idea what might have caused my mother's schizophrenia?"

After a deep sigh, Aunt Vermilla responded, addressing me by the given name I had left behind in Logan, "Oh, I don't know, Carol. I don't think anyone or anything is to blame; it was just a tragedy. Thank God it didn't happen to anyone else!"

Twenty-five years had passed since her husband, then a Mormon bishop, as well as the elected mayor of Bountiful, had declared my mother's marriage to a Gentile to be the source of her provocative female behavior. Had I been looking for pieces of my own life puzzle at that moment, I might have asked more questions and uncovered a few clues. Instead, I concentrated on letting go of my past, while preparing to create a new life.

In September, after the gravestone prohibited by my mother's will was in place, my father and I planted a lilac bush at the foot of her grave. It did not survive the winter. The following spring we planted marigolds and pink geraniums, incongruent with the plastic flowers preferred by other grave tenders. As her grave settled unevenly, the Louisville City Cemetery caretaker told us the settling occurred because the grave next to my mother's lacked a cement liner. My father's theory, always voiced as a declarative judgment, was different.

"The uneven settling is because you filled her grave yourself! You should have left that task to the proper authorities."

The last serious grave tending my father and I undertook was six years after my mother's death. As usual we drove in silence, stopping for eight bags of topsoil and a roll of sod. We dug up the old grass, added new soil to level the plot, and then put down the new sod in time for the fall rains. For many years after that leveling effort, I went to the cemetery alone. It seemed no matter the season, I always heard the song of a robin.

Eighteen years after my mother's death, Olivia and my first-born granddaughter, Tate, arrived from New York City for a visit. We stopped at the Louisville City Cemetery on our way back from the airport for the sake of three-year-old Tate, who skipped among the shaded monuments, trying to shed the four hours of energy she had stored on United Airlines.

Olivia took a photograph of Tate standing by her maternal great grandmother's gravestone. Looking at my granddaughter, a feeling of relief and gratitude washed over me that she would never know my mother. It was not until Olivia sent me the photograph that I noticed a strange coincidence. Tate was wearing an ice blue dress.

Then, like most mourners, I stopped visiting my mother's grave. My unrequited longing for a mother was finally gone.

Tate at the grave of her maternal great grandmother, Louisville City Cemetery, June 2001

-2-

My Son Is Not the Messiah

Three months after the death of my mother and the reversal of my tubal ligation, Bob and I began trying to conceive a child. I had one miscarriage in October, followed by Dr. Oliphant's prescribed pause. In December on Bob's forty-first birthday, my inherited instincts told me we had created a new life.

In the spring of 1983, during a rare appointment with Dr. Marquardt, he handed me a notice from one of his professional medical journals. Dr. John Nathaniel Rosen, who had nearly caused me to take my own life, along with that of my unborn daughter, had been stripped of all authority. Ironically, the downfall of Dr. Rosen was not due to the misogynistic psychiatric science he had advocated for so many decades. Instead, not only had he been caught lying about his professional training, but he had been physically and emotionally abusing patients at his Temple University Clinic near Philadelphia, as well as in his Florida facility.

On March 29, 1983, Dr. Rosen, the American Academy of Psychotherapy's 1971 Man of the Year, avoided being charged with sixty-seven violations of the Pennsylvania Medical Practices Act and thirty-five violations of the rules of the State Board of Medical Education by giving up his license to practice. The author of "The Perverse Mother had been felled by his own perverse ambitions and behavior.[1]

Because I was forty and hadn't given birth for twenty years, I was considered to be an "elderly *primigravida*" and underwent amniocentesis. The fetus was a male showing no chromosomal abnormalities.

My father's response to the test results, "Are you sure? Those tests aren't completely infallible are they? We don't know anything about raising a boy!"

The amniocentesis results meant that Bob and I had to decide what to do about a *brit milah* or circumcision. Bob had hoped to avoid this issue—so much so he had found himself wishing for a girl. Despite Bob's lack of religious practice, he did not want his second son to be the first uncircumcised male in his paternal and maternal bloodlines.

Bob was preoccupied with preparations for his first argument before the Supreme Court. I was overbooked with client obligations, so Olivia traveled from Rochester to hear her adoptive father as he stood before the nine Supreme Court Justices on April 27, 1983. But even after the argument and final briefs had been filed, Bob was still was struggling with the ritual circumcision issue.

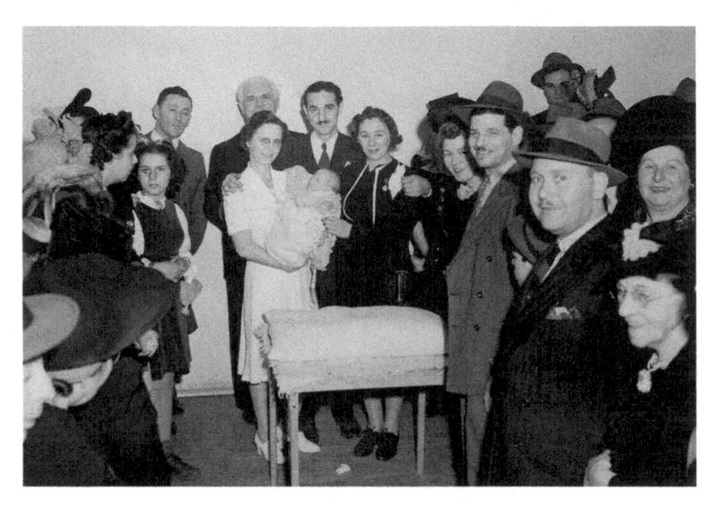

Brit Milah of Robert Stuart Pelcyger, Brooklyn, December 23, 1941

Bob at his bar mitzvah, Brooklyn, January 1955

As a child, Bob had gone to temple with his parents on Jewish high holy days and had studied for his *bar mitzvah*.[2] Having completed that passage, followed five years later by that of his brother, Joel, the Pelcyger family's temple attendance steadily waned. Bob recognized the contradiction, but his tribal roots and early cultural conditioning were fighting back. I understood exactly how he felt.

In August, I urged Bob to observe a *brit milah* at the home of Charlotte Smokler, our long-time neighbor who had become a close friend. It also provided him with an opportunity to meet the local *mohel*, a diamond cutter by profession. Bob felt faint during the cutting of the baby's foreskin. Afterwards the *mohel* said that unless I converted to Judaism, he would not perform the circumcision.

Then in my eighth month of pregnancy, even I knew it was not possible to adequately prepare for a conversion to Judaism in one month. Also, given my deep ambivalence about organized religion, fed by the high levels of protective pregnancy hormones then circulating in my blood, I found it extremely hard to think of either serious study or deliberately cutting anything on the fetus inside me. Bob and I finally reached a discomforting mutual decision that our son would not be circumcised. We had no choice but to live with the consequences of our exogamy and abandoned faiths.

Our son was born on September 20, 1983 at St. Luke's Hospital in Denver. After a long labor and full dilation, a caesarean was necessary to save his life because of pelvic disproportion (mine) and the circumference of his head (his). We named him Eben Valentine after his grandfathers. The "E" name was for his paternal inheritance: Bob's father, Eugene, who had a fatal heart attack in 1979. In Ashkenazi Jewish tradition, a child is not named after another person until that person is deceased. Eben is also short for Ebenezer, which means "stone of help" from the stone set up by Samuel after a signal defeat of the Philistines, as a memorial of the help received on that occasion from Jehovah. Valentine was for his confusing Gentile heritage.

As soon as the three of us were settled in a room, Bob's mother, Ruth, called from Florida with congratulations.

Ruth's first question to me was, "Where are you going to have the *bris*?"

Glaring at Bob as I handed him the phone, I gritted my teeth and whispered, "You didn't tell her?"

Bob sheepishly took the phone, stumbling to explain our decision to his mother. We three adults felt terrible, while Eben Valentine slept blissfully in my arms. A few hours after Ruth's telephone inquiry, the head pediatric resident from Children's Hospital came into the room.

"You need to postpone any arrangements for circumcision until you speak with the Chief of Pediatric Services!" he said, using a novice's stern attempt to exert medical authority.

Bob and I assured him we had no plans to circumcise Eben Valentine. Nonetheless, we found his message unsettling. Early the next morning, another resident told us a team of pediatricians from Children's Hospital had been up all night researching the condition of Eben's foreskin. Their conclusion was that Eben had been born with a natural circumcision—or, as the resident described it, "a natural circ."

When the Chief of Pediatrics arrived at my bedside, with the official news, he told us, "A natural circ is an extremely rare event—one in a billion!"

Bob and I laughed in joy and relief—a negotiated settlement must have been reached between his father and my mother in the no-man's land between their separate Gentile territories, heavenly or not.

Numerous doctors, residents, interns, and nurses came to see our son's "natural circ" for themselves. Bob and I had been culturally conditioned to circumcise or baptize a child into our world, but neither of us had the necessary faith to do so. In that sense, Eben Valentine Pelcyger was born free to find his preferred tribe and faith, or not.

On the third day of Eben Valentine's life, Bob drove us home from the hospital. As I carried our newborn son up our front walk, Moxie, an extraordinarily large male Golden Retriever belonging to our next door neighbor, Martha Hoover, lumbered toward us. I lowered Eben so Moxie's huge brown eyes could see what his canine nose had already detected. Bob opened our front door and I carried Eben Valentine inside his first home. When Bob went to retrieve my hospital bag from the car, he found the front screen door was blocked. Moxie had taken up residence on our front door mat and had to be coaxed to stand up.

For the next five years, Moxie rarely left our porch. At Eben's first sound each morning, Moxie's body would rattle the front door. If someone took Eben for a walk or brought him outside, Moxie was omnipresent. Whatever it was, Eben and Moxie had morphic communication[3] from the first moment their paths crossed. Moxie was religiously devoted to Eben for the remainder of his canine life.

Moxie died in June 1987, the same month I was invited to a dinner party for a Jewish Israeli scholar who was visiting the CU School of Law. The professor and I spoke about our families throughout the meal. He assumed I was Jewish from my surname. Over dessert I told him in a humorous mode about Eben having been born with a natural circumcision.

The face of the Israeli scholar suddenly lost color, "Ms. Lieberman, don't you know that Jews consider the birth of a child with a natural circumcision to be the sign of the Messiah!

"No," I responded, "I have never heard such a thing!"

"Would you mind if I came by your home to speak with him?" he asked. "I am very curious about what type of person he is!"

Suddenly feeling very protective of my son's normalcy, I responded, "Eben is a unique human being, but I can assure you my son is not the Messiah."

For the remainder of her life, Ruth seemed incapable of remembering her grandson Eben Valentine had been born with a natural circ, a sign of holy righteousness. Instead, Ruth was unable to forget she had already addressed announcements to her friends before being told that there would be no *brit milah*.

As for Moxie, Eben Valentine Pelcyger was his messiah and vice versa.

Martha Hoover with Moxie and Eben Valentine, October 1984

Moxie, Eben Valentine's favorite pillow, July 1985

-3-

Valentine and Orson

"Valentine and Orson, a tale good and rare,

Were two deserted infants suckled by a bear;

One remained all naked, in a sorry plight,

While lucky brother became a noble knight."

--An American Children's Book, ca. 1880

A lovey is a transitional object that provides comfort to a child when separated from the mother. D.W. Winnicott described them as objects which help a child move into the larger world, like a compass to negotiate the distance between self and other.[4] In one sense a lovey is also a child's first tool since all babies are born not realizing there is a separateness between themselves and their mother or their mother's breast.

Every child needs a lovey and only the child should make any alteration to it or determine when it is no longer needed. Most often a blanket or stuffed animal is transformed in this way, but sometimes a child uses a sound, a movement, or a body part like a thumb.

I had multiple loveys, including a blanket with satin binding, a gift from Aunt Alice. She purchased two identical blankets—one for her firstborn son, William Arthur Marshall (known as Bill) and the second for me.

When I was five, Alice pointed to Bill, who was holding his blankie, just as I was holding mine, and said, "Look, both of your blankets are just alike!"

Aunt Alice's gentle gesture, one which she intended to be a connecting bridge between first cousins, set off a panic button inside me.

I shouted, "No! My blankie is not like Bill's! It's different!"

Yet, I was ill-equipped to explain how my blankie was different in the face of arguments by my father and Aunt Alice. My only defense was that it was different because Nanie had sewn a replacement satin binding on it. I had sucked so hard on the original, it had dissolved. I had no idea at that age why I felt so desperate to defend the uniqueness of my transitional object, but I felt certain that my source of comfort was singular, not a twin.

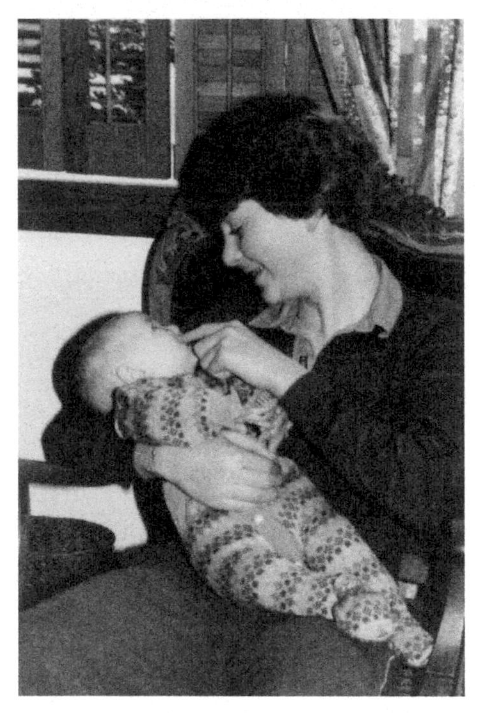

Jordan rocking Eben in platform rocker, December 1983

In addition to my blankie, I often fell asleep to the sound of squeaking springs on an old platform rocker, one purchased by my parents at a Delta estate sale. The platform rocker has survived. It has been re-upholstered five times and refinished twice, having been moved through the serial homes of my parents until it became part of my own Idaho household. Having rocked Olivia and Eben in it, I offered it to my daughter when she was pregnant with her first child.

Olivia politely declined my offer, saying, “No, thank you; it squeaks.”

Teda was another lovey, one vitally important to me from birth to age eight. Nanie created Teda for me, making him come alive with bright button eyes and an embroidered smile. She covered his head and belly in recycled nubby brown wool, his four chubby limbs with striped seersucker cotton.

In 1950, my mother discarded my beloved Teda so he would not "contaminate" her new Dream House. When I discovered Teda was gone, I was beyond consolation. My father took me to the Logan City Dump to search for him, but Teda was not to be found. I never forgave my mother for that cruel act. The thought of my beloved friend lying cold and wet amongst the rubble before being burned up haunted me for years. It was my own personal variation of the well-known children's story, *The Velveteen Rabbit (or How Toys Become Real).* Something would remind me of Teda and I would be blind with tears before I even recognized why I was weeping. That loss was the last time I ever let my mother know what was important to me. From that moment on, I tried to keep whatever I cherished a secret from her.

Two months before the birth of Eben Valentine, Gale sent her half-brother-to-be a pair of booties in a small gift box. On each of the six sides of the box was an illustration of a different fairytale. One of those six, *Valentine and Orson,* caught my eye because Bob and I had just decided Valentine would be Eben's middle name. I went to the library where I learned *Valentine and Orson*[5] is a French fairytale about twin brothers. The earliest known version was written in verse in 1489.[6]

Gale's gift box to the author, July 1983

According to the fairytale, Valentine and Orson were the twin sons of Bellisant. She was the sister of Pepin, the King of France. She was married to Alexander, Emperor of Greece. In a fit of jealous anger, Alexander banished Bellisant. Retreating to France, she gave birth to their twin sons in a forest near Orleans. The infant Orson (French *ourson*, 'little bear'), was carried off and suckled by a bear. His twin brother, Valentine, was rescued by his uncle, King Pepin. When their paths crossed as grown men, Valentine saved Orson's life.

Remembering Teda, I left the library and drove directly to Grand Rabbits Toy Shoppe in search of a teddy bear for my unborn son. I named him Orson, hoping he would become Eben Valentine's lovey. And he did.

Orson and Eben Valentine snuggling 1985

Eben Valentine pulling Orson, April 1985

-4-

Adieu Manx and Grey Cat

The same year Eben was born, Grey Cat celebrated her sixteenth birthday; her son Manx, his fourteenth. Grey Cat had lived a long life for an Idaho barn cat, despite a major health crisis the year Olivia graduated from high school. Having become extremely thin; Grey Cat was finally diagnosed thyroid cancer.

After chemotherapy and surgery to remove her thyroid, Grey Cat was fine on replacement medication for several years. During that time, her only health problem was that she had lived so long her teeth were worn down to her gums. Raised on Idaho beef liver, Grey Cat made her preference known by vocalizing her objections to any other offering. Absent any teeth, Grey Cat's meals of raw liver had to be pre-shredded before she could gum it down, a chore our children sought to avoid with increasing creativity.

During the early months of my pregnancy, Manx had also begun showing symptoms of thyroid cancer. Manx suffered a complication during pre-surgical chemotherapy necessitating a two-week hospital stay at the College of Veterinary Medicine in Fort Collins.

Manx after chemotherapy and surgery being nursed by Grey Cat, March 1983

Manx eventually recovered and was fine—until Eben Valentine was born. To Manx, Eben was an interloper, who interfered with his snuggling time with Bob. Manx soon started an insidious campaign of feline revenge.

Eben was a month old, when one of my clients flew to Colorado from Florida to work on a speech she was scheduled to deliver. Ready to begin dictating text for the ideas we had been discussing, my client popped a tape out of my Dictaphone transcription machine, only to discover it was dripping in malodorous male cat urine. The Dictaphone representative laughed at our request for repair service.

The Dictaphone was the most costly, but not the most egregious, of Manx's many targets. In desperation, we became the first cat "guardians"* in Boulder to hire the equivalent of a feline psychiatrist. An animal behaviorist from the College of Veterinary Medicine arrived to assess Manx's psychological state.[7] In the misogynistic tradition of Dr. John Nathaniel Rosen, the therapist's first questions were about Manx's mother.

"Did the mother show any cruelty to her litter? "How did Manx interact with his litter mates?"

Manx unhappy to have been displaced from Bob's lap by Eben

* The City of Boulder adopted an ordinance making owners of pet animals "pet guardians."

Newspaper photo from Colorado Daily, November 24, 1983

Manx's psychotherapy was so unusual the *Colorado Daily* carried a story about it, accompanied by a photograph of Manx. Which is more shameful? That we hired a cat therapist? Or that we allowed the newspaper to do a story that identified us by name?[8]

With considerable reluctance, we finally determined Manx's rivalry with Eben could not be mediated. We persuaded our veterinarian to come to the house because Manx was terrified of the animal hospital. On Bob's lap, Manx had one last cuddling session ending with a nose-to-nose kiss from Bob before we nodded permission for the veterinarian to put him to death. We buried Manx in the garden graveyard, beneath the cherry tree by Olivia's bedroom window, next to his sister, Red.

Grey Cat lived for another year, then her kidneys began to fail. The day after Thanksgiving 1984, with Olivia cuddling Grey Cat in her arms, I drove the three of us to Alpine Animal Hospital. Then, sobbing in sync, we brought Grey Cat's body home. We buried her next to her two children, Manx and Red. Grey Cat was the finest and last feline of my life.

-5-

Falling in Love Again

Eben's birth came twenty years after Olivia's. I had forgotten that having a baby is like falling in love again. I moved into late life milky motherhood, a slow swimmer in the deep sea of maternity where I was both everything and nothing to another human being.

Instead of being preoccupied with the presence of Olivia and Bob, I focused on my newborn son, laughing at myself for giving the kind of attention to his wardrobe that Mildred Verhaag might have provided. For my last appointment with Dr. Marquardt, I carried Eben Valentine in my arms—my own therapeutic version of show and tell.

The author with Eben Valentine, Santa Monica, June 1984

As Eben Valentine moved into toddlerhood, I saw how I was recreating a childhood for myself, as well as making one for my son. Anything that interested him, also interested me—from sand and water to seed pods and butterflies to watching clouds and rolling on the grass. I felt as if being fully alive was a secret that had been kept from me.

Frank V. Lieberman also experienced a kind of rebirth after purchasing a new nest in the Friends' Condominiums—a fourteen-unit project developed by five couples, all long-time friends, who were professors at CU or government scientists.[9] The founding friends wanted to live near each other during their retirement years. Instead of a swimming pool, they chose to have a common library. This was perfect for my father since he couldn't even float in the Great Salt Lake. His new condominium was six short blocks from our home on Marine Street. At his request, I helped him shop for new furnishings, including a luxurious full-length couch upon which he took many naps.

My father continued to come to our home almost every day, but now his main purpose was to see Eben. The two Valentines had an instant affinity. Eben's first word was to reach out and call for his "BaBa." After that, in our efforts to support Eben's language development, my father was rarely addressed by any other name.

Olivia was now a doctoral student in clinical psychology at Adelphi University on Long Island, having graduated with honors like her adopted father from the University of Rochester. Before beginning her graduate studies, Olivia had studied at the Tavistock Institute[10] in London and traveled through Europe. A few days after Olivia's return, we staged a garden party (our first catered event) in honor of her twenty-first birthday. Beautiful in every way, she continued to have an impressive level of emotional intelligence.

Eben Valentine with his BaBa, Boulder, December 1984

While my two progeny were thriving, I was developmentally out of sync. My friends no longer had young children; my clients and colleagues were either perplexed or annoyed by my new focus. While I was nursing, I limited almost all my leadership and management consulting assignments to those allowing me to return home at night. In addition to writing articles for professional publications, I contributed a chapter to a book edited by Larry Hirschhorn, a Wharton colleague.[11]

My consulting work was increasingly with businesses or organizations led by women and I was speaking about female leadership at conferences and retreats. Those presentations led to a contract from a university press for a book entitled *Gender Anxiety* focusing on the behavioral changes in subordinates when a woman is at the top of the hierarchy.

I used the fifteen hundred dollar advance to purchase an antique couch without consulting Bob. Although I eventually returned the advance, for the next decade Bob refused to sit on that couch as his way of expressing appropriate annoyance for my having treated the publisher's advance as "my" money versus "our" money. Ironically, Bob's couch behavior replicated my mother's refusal to sit on the Dinwoody's couch she had imagined was filled with poisonous snakes, as well as the couch détente between Ben and Afton Evans.

Hulett (Bucky) Askew holding Eben Valentine and handmade music box, February 1984. "Gender Anxiety" couch in background.

Several clients came to Colorado to work with me. One of my favorite clients, Hulett (Bucky) Askew, arrived from Washington, D.C. with a handmade music box he had crafted for Eben, who ignored the box and instead fixated on Bucky's beard.

I was working extremely hard so we could help with Olivia's graduate school expenses and Gale's undergraduate tuition at Wesleyan, while also trying to ameliorate Bob's concern about our financial resources. Jordan, who was now living with us full time while attending Boulder High School, was a great big brother. However, I couldn't have managed Eben's care, while also continuing to work as a consultant, without my father's support. He was the best "Manny" in the world.

Jordan and Eben, big and little brothers, January 1985

In search of peers for Eben, I organized a playgroup. At a toddler's gym class, I met Hope Morrissett and Bev Cole, whose sons, Nicholas Gold and Cole Davis, were born within a few weeks of Eben. Although both mothers were ten years my junior, we began gathering for playtime, field trips, and motherhood talk. Soon two other mothers of daughters of the same age joined us: Caryn Small, her daughter, Rachel, along with Carol Hartman, and her daughter, Cassidy.

As the five children approached their third birthdays, we began to meet less frequently because they were old enough for formal preschool experiences. Eben began attending New Horizons Preschool, a parent-teacher cooperative, founded in the aftermath of the assassination of Dr. Martin Luther King. New Horizons was located in the Friends' Meeting house—the same building where twenty years earlier Olivia had re-started her elementary education at Upland School.

A few weeks after Eben's third birthday, I had my annual OB-GYN checkup. As I drove into Denver, I remember singing along with Dionne Warwick, "That's What Friends Are For." I felt happy and confident about the future. Dr. Oliphant was running late, a hazard of his specialty. When he finally finished his thorough physical, he sent me across the street to St. Luke's Hospital for my first mammogram. As a mammogram novice, I felt no worry when the technician asked me to wait. I was still sanguine when she told me the radiologist wanted some "additional impressions."

Returning home two hours behind schedule, the telephone was ringing as I opened the front door.

Dr. Oliphant said, "Joan, I am very sorry to tell you that your mammogram shows several areas of significant calcifications."

Speaking only to myself, I thought, "*Doctor, of course, I have calcifications! I am producing a gallon of calcium-rich milk every day*!"

Dr. Oliphant interrupted my silent soliloquy, "We need to biopsy these areas. I have asked Marilyn to get you the first available appointment with R. Lee Jennings. She will send your films over to his office. He is the best breast tumor surgeon in Denver."

I started to ask a question, but Dr. Oliphant interrupted, "Joan, please excuse me; I am being paged for an emergency. We can talk further when you come in. Marilyn will call you as soon as everything is set up."

When I returned to Denver, Dr. Jennings told me I had a new challenge, one as life-threatening to my mothering of Eben as my fear of schizophrenia had been to my mothering of Olivia. I had breast cancer.

-6-

Friends' School and the Help Family

The day before Halloween 1986, two malignant tumors were removed from my right breast. Flexing my super-mom muscles, I staged a previously planned Halloween party for the families of Eben's play group. I was forty-four years old. Until the week before the surgery, I was still breastfeeding. Both Eben and I accepted our suddenly imposed abstinence with grace.[12]

Addie Help came into my life the same week I learned I had breast cancer. When I first encountered Addie, she was sobbing, while facing the back side of her coat cubby at New Horizons Cooperative Preschool. Addie's long, almost white blond hair was damp with tears and her pale blue eyes were fringed in white lashes—features reminiscent of Snowball. Addie was three and one-half years old, but tall enough to have passed for a five-year-old. Her too-short blue jeans revealed unmatched socks.

Trying to distract Addie from her distress, I sat down on the floor in front of the adjoining coat cubby and, blindly reaching behind me, grabbed a book from a stack on top of the cubbies.

"Hello, I'm Joan, Eben's Mom," I said, pointing to Eben, who was noisily stacking blocks in the main room. "I thought I might sit here while I read a book. Would that be okay with you?"

Without looking at me, Addie nodded permission. By the time I finished reading the first page, Addie was sitting next to me. The book was Margaret Wise Brown's *The Dead Bird.*[13] In those first moments of interaction, I had no idea the shadow of death would define my relationship with Addie Help.

Addie slowly showed signs of comfort in my presence—we read her favorite books, fed the school animals, and sat next to each other at the snack table and during closing circle. It took very little interaction with another human being to trigger Addie's silent tears. Almost any comment, attention, or pressure produced the same result. On the rare occasions when Addie spoke, her voice was so faint others had to strain to hear, but her animation doubled around the guinea pigs, the salamander, the fish, the rabbits, and chickens. There was one human exception: Addie's mother. Whenever she arrived, Addie broke into a wide smile and skipped to her side.

I was afraid of dying, but more terrified of what my death would do to Eben. Watching the anonymity created by the size of the New Horizons class, I persuaded the other mothers in the original playgroup to supplement the larger, more anonymous preschools our five children were now attending by reconstituting the old playgroup. I wanted a loving community of substitute mothers for Eben in the event I didn't survive.

Taking the lead, I agreed to create a formal preschool classroom in my consulting office on the second floor of our home. We hired a teacher (Cindy Uhr) and an aide (Ottila Hernandez), and planned for a June opening.

The author's consulting office in November 1985 with drafts of "Gender Anxiety" book and ice cream chair from the Inland Hotel in Troy, Idaho, before it became a classroom for Friends' Preschool.

The make-up of the old play group had been three boys and two girls. To improve both the budget and the gender balance, we needed another girl. Bev Cole suggested Addie Help because she had heard Addie's mother, Jane, had just been diagnosed with breast cancer. As Bev spoke, I remembered seeing Addie's mother nursing a small baby in the play yard of New Horizons. She was tall and slender, with short strawberry blond hair and legs that still looked unbelievably great in shorts.

I wondered whether Bev's information was correct because both my tumor surgeon and my oncologist had told me to stop nursing immediately. Yet, I was drawn to the idea of a cohort, another mother with young children, unexpectedly facing her mortality. In retrospect, this was a very small need of mine. I have often wondered how different my life would have been had I ignored the feeling of isolation and stigma that accompany a cancer diagnosis. But any inclination to look for another child, or to think more deeply about having two mothers with breast cancer in such a small group, was lost as soon as I imagined Addie making her way up our staircase to a place of safety.

The addition of Addie to our fledgling preschool turned out to help gender balance, but not the budget. When we learned the Help family had eight children, we asked only a token tuition and recruited a fourth girl for budgetary purposes, Lindsey Stuart, the daughter of Janie and Terry Stuart.

Left to right: Eben Valentine, Nicholas Gold, Cassidy Hartman, Lindsey Stuart, Ottila Hernandez, teacher's aide, and Cole Davis.

First class of Friends' Preschool, clockwise at rear: Eben Valentine, Nicholas Gold, Cassidy Hartman, Addie Help, Cole Davis, Rachel Small, and Lindsey Stuart, standing. Thanksgiving, 1987

On a cool April morning, Jane and I stood talking on the sunny side of the family's battered Chevrolet van as her eight-month-old son, Adam, reached inquisitively for one of my earrings. Jane told me her breast cancer had been diagnosed two weeks before Adam's August birth. After his birth, a surgeon performed a lumpectomy. Before leaving the hospital, Jane was told that under no circumstance was she to nurse her eighth baby. Not only was Jane told not to breast feed, but because her tumor was particularly aggressive she was advised to immediately begin chemotherapy, to be followed by radiation.

Jane ignored all those warnings. She had been breast feeding babies for nineteen years and had been the leader of the local chapter of La Leche[14] for almost a decade. Nursing was central to her life. Perhaps it was not even a conscious decision to forego treatment—maybe Jane was simply drunk on the standard hormonal cocktail of oxytocin and prolactin produced by pregnancy and nursing. Whatever it was, Jane lacked any sense of urgency. Adam, like billions of other babies, would have been fine on formula, but he would not be fine without his mother. Nor would his siblings. Ignoring the real risks of producing breast milk, Jane stayed in the world of mother and infant, surrounded by the protective instincts all female species have for their newborns.

When the children began gathering in my former office, Addie found her own voice almost immediately. A group of six or seven was a size Addie knew intimately from her own family, in contrast to the frequent chaos of twenty children at New Horizons. The seven children named their new preschool "Friends' Preschool" because it was where they went to be with their friends. More than one of them began asking their mothers: "Is today the day I go to my friends' school?" With their teacher Cindy Uhr's superb sense of celebration, the name of Friends' Preschool was soon made official.

KID AND DINOSAUR READING
by
Eben Valentine Pelcyger

A little kid went up the mountain with his BaBa, and do you know what the little kid's name was? Eben. And the BaBa's name was Frank. They went up the mountain and the little boy chopped the rocky earth with his hatchet. A neck sticked out of the mountain. The kid asked BaBa, "What is that sticking out of the mountain?" BaBa said to the little kid: "It's a dinosaur neck and its body is inside the mountain." Then the little boy chopped up every single rock so the dinosaur could get out of it, and they carried it home in the dump truck. And do you know where they keeped it? In the back of the dump truck, and they drove it home to the mother. "What's that?" said the mother to the little boy with the hatchet. "It's a dinosaur and we are going to keep it forever."

"Remember," he said, "it is a leaf eating dinosaur."

Eben Valentine's first dictated story, recorded by Cindy Uhr, Friends' Preschool Teacher, 751 Marine Street, Boulder, Colorado, February 22, 1987

My contact with Jane during the first year of Friends' Preschool was minimal. Her aura of calm caused me to hide my cancer anxiety in her presence. Instead, Jane and her family became an important reference point, providing me with an alternate reality. As my fatigue increased in proportion to Eben's exuberant energy, I would try to quell it by imagining the demands Jane was facing from eight children. Every time the mother in me felt a little tired or overwhelmed, I would simply multiply the task by eight and continue my work.

Addie's parents, Jane and Tom Help, having met as students at the University of Colorado, built a family and a business in Boulder. During my first conversation with Jane, I was surprised to learn that they were neither Catholic, nor Mormon. For reasons Jane never fully explained, they had created their own congregation.

Cancer made social intercourse more exhausting for me. Suddenly the concerns and needs of others were harder to hear, as if there was background static from an annoying radio station. Feeling insidiously distracted, I wondered how Jane was able to listen to, let alone remember and act upon, the needs, wishes, and worries of eight children.

In August, the Help family went to Yellowstone. It was late September when Jane told me that she had been "pretty uncomfortable" while hiking with Adam in a backpack, who was now twelve months old. After telling me that she hoped Adam would start walking soon, Jane casually mentioned that Dr. Julie Carpenter, the Help's family doctor, had persuaded her to have radiation to the affected breast. She had been making the daily trip to St. Luke's Hospital in Denver for a week. Boulder was then still a medical boondocks, with only three oncologists and no radiation treatment center, although one was then under construction.

As soon as Jane told me about her weekday treks to Denver (all treatment schedules seem to be based on the premise that cancer cells don't grow on Saturday or Sunday), I offered to help with Addie's transportation. That simple offer was all it took—without even knowing it, I bought a one-way ticket into what I came to think of as "Help World."

The Helps lived in a subdivision in the northeastern corner of Boulder. The house was a bi-level with a garage door facing the street. The three small bedrooms and a full bath on the second floor were occupied by their six daughters. Jane's husband, Tom, had converted the unfinished basement into a master suite with a bathroom and another small bedroom separated by a laundry alcove. The small bedroom was occupied by their twelve-year-old son, Ronald. Adam was still sleeping with his parents in their king-size water bed. In the tradition of suburban America, their double garage sheltered no vehicles. Rather, it was overflowing with tools, bicycles, and other developmental accessories haphazardly acquired by American families with children.

It was several months before all the faces and names of the Help children came into focus. The three eldest girls, Christina, Jamie, and Tara, were in the mold of America's female icons—tall and slender, with long blonde locks. The fourth-born daughter, Lizzie, bore a stronger resemblance to Tom. Ronald had been the only male in a sea of six sisters until Adam's birth, but he seemed to have found ways to stand apart without evidencing any of the edginess that can come with gender isolation. After Ronald came Katherine, then Addie, and finally baby Adam.

The Help children appeared to me to be unusually polite and respectful. I never observed a display of temper or sulkiness; they each responded to any parental request, seemingly without hesitation or resistance. I wondered whether these character traits were genetic, a result of great parenting, or a basic necessity in a large family system.

Katherine was as soft spoken and gentle as Addie and they were emotionally entwined. I soon realized that if I was going to invite Addie for an outing or a meal, Katherine should be included as well. While Addie had a strong preference for animals, Katherine radiated both her appreciation of and longing for contact by being acutely sensitive to the needs of others. Unlike Addie, but like Eben, Katherine suffered from severe asthma. Our knowledge of how to manage asthma made our home an especially safe place for Katherine. It was not long before our house felt empty without both of them.

During the second year of Friends' Preschool, I persuaded Hope Morrissett and Bev Cole to join me in an effort to continue Friends' School into the primary grades. In addition to my own mortality challenges, I was also motivated by the memory of Olivia's experiences in Troy. After many years of debate and discussion, the National Association for the Education of Young Children (NAEYC) had finally established a formal set of standards for early childhood education programs. In addition to a child-centered curriculum, those standards called for two teachers for every group of fifteen children. Having been involved in the early drafting of the standards, I was well aware there were no schools in Boulder, either public or private, which met them. Still stuck in my 1960's idealism, I thought we should try to create a model of the ideal using Friends' Primary School, one that would meet NAEYC's standards, lest the world forget what is optimal.

Katherine and Adam Help, Marine Street, May 1988

Since my work with Head Start, whenever elected government leaders begin proposing cuts in educational funding or the revision of curricula, I have wanted to insist that each has to first assume sole responsibility for managing a classroom of twenty or more young children for one week.

I started scouting Boulder for a space that would meet both licensing and zoning requirements. In the end, the solution was a small rental house on the east side of our home on Marine Street. We were able to obtain a year-long lease with an option to renew after promising to pay the tough-minded owner in blood if the children did any damage to her property.

In May 1988, we staged an overly elaborate ceremony for the Friends' Preschool graduates. One of my favorite photographs is of the seven graduates with their mothers standing behind them. Two of the mothers were actively dying, but no stranger could have guessed which ones were ill. None of us had yet allowed ourselves to imagine any of our children would also be stricken by cancer, but two were and one did not survive. Of the seven mothers, only two remained cancer-free over the next ten years.

First Class of Friends' Preschool, Graduation Day, May 1988, Boulder, Colorado. Children and Mothers (left to right): Lindsey and Janie Stuart, Nicholas Gold and Hope Morrissett, Eben Valentine Pelcyger and Joan Carol Lieberman, Cassidy and Carol Hartman, Rachel and Caryn Small, Addie and Jane Help, Cole Davis and Beverly Cole.

-7-

A Puli Named Marley

A puppy was the only gift Eben wanted for his fifth birthday. The death of Moxie, his canine guardian, hit Eben hard.

"I feel lonely, Mom." he repeatedly told me. "I need a dog."

This was not a simple wish. We had to find a "non-allergenic" canine because Eben had developed asthma. There isn't really such a thing as a non-allergenic animal, but we convinced ourselves that Eben could tolerate living with a dog that didn't shed.

My dental hygienist, Andra Beach, had spent months looking for such a dog, finally settling on a Puli, a Hungarian sheep dog. Her Puli, Puck, became our shopping model. Pulik (plural) have hair, not fur. The hair grows into dread locks so that Pulik resemble the sheep they instinctively guard. While I don't believe, as some do, that Pulik were in the Garden of Eden, the breed is one of the oldest.

On his fifth birthday, Eben awoke to a dog kennel, but no puppy. Inside the kennel was a book with a note from the "birthday fairies"—a reminder that some remnants of Utah-style magical thinking were still operational in his mother's brain.

Dear Eben,

A female puppy named Marley will be coming to live with you in ten days. Please have your parents read this book to you so that you will be ready to take good care of Marley when she arrives. We are certain she is going to be a great canine friend, just like Moxie.

Love, Your Birthday Fairies

Marley was located through the resourcefulness of a breeder in Mead, Colorado, whose male Puli had helped produce a single female pup. The female puppy had been named Marley because she was born on the birthday of the Rastafarian musician Bob Marley. It was necessary for her to be spayed before we were allowed to adopt her, hence the delay. The breeder, also a United Airlines flight attendant, kindly arranged for Marley's flight from Omaha to Denver.

On the last day of September, I drove Bob and Eben to Stapleton Airport. After dropping Bob at the main terminal for his flight to Reno, Eben and I went to the Freight Center to wait for Marley. As soon as we sat down, Eben began asking in thirty second intervals when Marley would arrive.

His expressions of impatience were followed by a question, "Mom, do you think Marley will like me like Moxie liked me?"

When Marley's baggage cart was pushed into view, we slowly bent down to peer inside her kennel.

Eben's soft little boy voice called, "Marley, are you okay?"

A pair of dark brown eyes surrounded by a mass of black fluff came to the opening and met our own blue ones. Eben carefully attached a new purple leash to Marley's collar and took her outside to use the grass strip along Quebec Boulevard. While Eben gave Marley a drink of water, I chose bonding over safety, storing Eben's car seat in the trunk so that he could sit right next to his new canine companion. I watched through the rearview mirror as Marley put her head on Eben's thigh and snuggled close. My five-year-old son seemed paralyzed with pleasure. Before we reached the turnoff to US-36W, both dog and boy were drifting into sleep mode.

When I parked in front of our house, Marley sat up, while her human seat-mate remained sound asleep. I led Marley through the side gate to our backyard fishpond and garden. While she and I were exploring, my father arrived. Spotting his sleeping grandson, my father went to check on him. Marley immediately began barking and straining the leash to return to the car. When my father realized Marley was already protecting Eben from danger, he immediately fell under her Puli spell.

The next day I realized Marley was going to be my Moxie. After Eben left for school, Marley slowly inched herself across the floor to where I was sitting and put her chin on my foot. She entered my psyche, in the same way Moxie had entered Eben's. She rarely left my side—a hike with my father was one of the few exceptions. We anticipated and understood the needs of the other without speaking. When she sensed I was tired, Marley would jump up on our bed or chaise to signal it was time to rest. I felt her hunger as my own; Marley seemed to know my worries before I did.

Eben Valentine with Marley, his Puli, November 1988

When hiking with my father, Marley never lagged behind, went as far and as fast as he wanted, and always remembered the route home perfectly. My father worshipped silence, so Marley quickly became my father's perfect hiking companion.

He often said, "Marley is quiet—even though she is a female, she is not like all those other babbling ones."

From the beginning Marley accompanied us everywhere. With the command to stay, Marley patiently waited without being leashed or allowing herself to be distracted. Once my father left her at the side entrance of his pet-free building. He went up to his second floor condominium to change clothes, exited the building through the lobby door to the garage, completely forgetting that Marley was waiting at the door on the other side of the lobby. He drove out of the garage, intent on doing several errands. Returning four hours later, Marley was still waiting exactly where he had left her.

Professor Lawrence Senesh and his wife Dorothy Marchus were among the original founding couples of the Friends' Condominiums. Professor Senesh, who was born in Hungary, told us that the first Pulik brought to America had been bred by his family. They were transported across the Atlantic in 1939 by Professor Senesh's nephew, Dr. J. Eichorn. Professor Senesh and Dorothy raised two of the first litters of Pulik born in America and subsequently became devoted parents to three generations of the breed. They had no human children, only Pulik. We often brought Marley to visit Professor Senesh, who would greet her with absolute delight on the front lawn.

Dr. J. Einhorn bringing the first Pulik to America in 1939. They were bred by the family of Dr. Lawrence Senesh in Hungary.

One day we were out walking in a rural area north of Boulder, when we came upon a small herd of sheep guarded by a llama. The llama ran up and down the fence line in tandem with Marley. Then in a momentary pause, the llama used his height advantage to spit a giant gob of mucus into Marley's face. Stunned, but undeterred, Marley found a place to squeeze under the fence, quickly taking control of the whole group, llama included. My memory of her performance that day still makes me smile. Whenever we were near that area, Marley urged us to return to her favorite herd for another visit.

Marley herding Llama and Sheep, Oxford Road, Boulder County, 1992

-8-

Acts of Sublimation

In the first class of Friends' Primary School, three of the mothers of the fifteen kindergarteners had breast cancer. The disease suddenly seemed epidemic; its origins mysteriously multi-faceted.[15]

Besides Jane and me, the third was Margaret Martin Clifford. Margaret, then forty-four, was married to Glen Clifford, a family therapist. They had two children: a son in middle-school, and Katie, who was four. Technically, Katie was too young for the first kindergarten class at Friends' School, but we admitted her because Margaret had cancer. Also, Katie brought a sliver of diversity to the class since she was Mexican by birth, adopted when she was six weeks old.

What I remember from my first meeting with Margaret was her strong physical resemblance to Jane. Both were tall, slender strawberry blonds, with coloring that made them appear backlit by sunlight, even in the evening.

After Addie missed a week of school in early December 1988, I telephoned the Help's home. Jane's third-born daughter, Tara, answered.

In a whispered voice, Tara said her mother wasn't feeling well, adding, "Mrs. Lieberman, could you come over, please!"

Not only had Jane's breast cancer spread, the tumor was wrapped tightly around her spinal cord. She was in agony. In their collective denial, the family had waited until Jane literally couldn't crawl up the stairs from their basement bedroom. It took four men to maneuver her off the water bed before gingerly hoisting her body up the narrow stairway to the waiting ambulance for the short trip to Boulder Community Hospital's Mapleton Center.

Fortunately, Boulder's new Miriam Hart Radiation Center had just opened. Every day an ambulance crew transported Jane from the Mapleton Center five blocks east to the Radiation Center. The director was the same radiologist who had treated Jane's left breast in Denver. He told me he was being very careful because Jane had stopped coming for treatment at St. Luke's Hospital after only four of the recommended six weeks of radiation.

It was close to Christmas. I started bringing a few meals to the Help home and invited Addie and Katherine to join our holiday activities. I tried to pay short daily visits to Jane during the hours our children were in school. Even in pain, she maintained a kind of regal calm. I brought useless gifts. What Jane really wanted was something Tom brought—a large photograph of all the kids and himself in full color which he pinned on a cork board close to Jane's bed.

Either Tom or one of their older daughters brought Adam to the hospital several times each day so Jane could continue nursing him from her unaffected breast. I have no idea what the doctors or the nursing staff thought; most likely they didn't know. Nor do I know what consideration, if any, was given to the effect on Adam of either the radiation or the morphine circulating in Jane's body.

The radiation eventually gave Jane enough pain relief that she was able to stand with the help of a walker. She was discharged from the hospital the second week of January, still very weak and still nursing Adam. Dr. S, then one of Boulder's three oncologists, recommended Jane start chemotherapy as soon as her white blood counts were closer to normal range.

After Dr. S made this recommendation, Jane and Tom decided they wanted a second opinion. I was surprised when Tom asked me about my oncologist in Denver. I felt a strange panic in the presence of what seemed like Jane's total passivity about saving her own life. Yet Jane acted as if she expected to survive. I attached myself to that vibration, absorbed the wish, and swallowed the bait.

Besides a few brief conversations in the hallway outside Jane's hospital room, the trip to and from Denver was the first sustained contact I had with Tom. At his request, I chauffeured the couple to Dr. T's office at the Mid-Town Medical Center. Jane rode in the back seat of our sedan surrounded by bump-buffering pillows. I drove gingerly, through an endless maze of winter potholes. Tom sat in the front passenger seat; Jane's wheelchair was stowed in the trunk. I tried to engage Tom with questions about his business and children. His barely audible responses came with minimal detail. Blond and blue-eyed like his children, he stood just under six feet. Significantly overweight, Tom seemed ill at ease, his body mass as dense as his social defenses.

Dr. T hung Jane's films on the back-lit screen in his office—the images showed extensive metastatic disease. Dr. T told Jane he could not cure her.

"But," he said, "I can extend your life by some months, provided we give it both guns!"

Dr. T's plan of attack would be to give Jane very high doses of chemotherapy, as well as more of the radiation she had just completed. From my own appointments with him, I was already familiar with the plethora of military aphorisms he used. Having come home from the War in Vietnam to join the War on Cancer, Dr. T habitually used the lingo of the Mekong Delta to describe his medical arsenal.

The menacing finality of Dr. T's prognosis could no longer be ignored by Jane. Still I was surprised when she agreed to start chemotherapy that afternoon because until that moment she had not even agreed to Dr. S's less aggressive plan to start when her blood counts were closer to normal. Jane was wheeled down the hall to the chemotherapy area, turning away the company offered by both her husband and friend.

While we waited, Tom and I walked around the grounds of the Mid-Town Medical Center. Only after both of our bodies were thoroughly chilled did we return to the tense waiting room. We sat side-by-side, Tom thumbing through old magazines, me writing in my diary, until Jane was wheeled out, pale, shaking, and permanently altered.

We were three hours late when we started the forty mile drive back to Boulder, a drive made more interminable by our fear and emotional exhaustion. Jane rode prone on the back seat, heavily drugged with Compazine to control nausea. Tom and I barely spoke; there was too much to say.

It was dark when we pulled into the Help driveway. I got out to retrieve the wheelchair while Tom roused Jane. We struggled to maneuver her off the backseat. Then Tom opened the garage door, roughly clearing a path through the clutter, before pushing his wife to the door leading into the house. Twelve-year-old Ronald opened it while holding Jane's walker. She was helped to stand, and, after a long pause, began to move forward.

Jane's close friend, Susan Eikenbary, had been caring for Adam all day. As Jane painfully made her way into the living room, Susan moved toward Jane, holding Adam in front of her, and began giving Jane maternal cues.

"Adam, look! Here's your Mommy!" Susan said. "Now you can nurse and get that nap you didn't take this afternoon!"

Instead of reaching for Adam, Jane uttered one sharp, long wail. Tortuous time passed, during which all sound seemed gone from the world.

Then Adam began screaming, "Mommy! Mommy!"

Jane teetered and shifted all of her weight to her left arm before slowly using her right one to push on Adam's stomach, as if trying to put him back into Susan's arms and care.

Looking past them both, toward a full moon rising above the back yard fence, Jane wailed, "I will never be able to do that again!"

Only Addie, Katherine, and Ronald were at home—like deer caught in headlights, they froze in place. Tom finally intervened and took Adam from Susan, carrying his son through the sliding glass doors into the backyard.

Adam bawled, "Mommy, I want Mommy!" as his arms struggled to reach for his mother over Tom's shoulder.

There was no privacy for Jane; her hospital bed was in the center of the living room. Susan and I helped her onto the bed. Then Susan made a hasty departure, wiping away her tears as she fled. I filled Adam's baby bottle with milk from a refrigerator extremely close to empty and put the bottle in a pan of water on the stove. Then I asked the three stunned children if they had eaten dinner. When they acknowledged they had not, I suggested that the girls spend the night at our house and that we get hamburgers and milk shakes at Good Times before we dropped Ronald at Centennial Middle School for his league basketball practice. Readily agreeing, they quickly dispersed to gather their things.

After putting two frozen TV dinners into the oven, I took the warm bottle and one of Jane's homemade quilts outside to Tom and Adam in the cool moonlight. When I came back inside, Jane was snoring and Addie, Katherine, and Ronald were already in my car, anxious to make their escape.

Two days after that wintery February evening, Jane was readmitted to the Mapleton Center, her white blood cells barely measurable. One day when Dr. T was filling my own veins with the same toxic cocktail of Adriamycin, FU-5, and Cytoxan, (CAF), he responded to my question about the efficacy of the dose he had given Jane.

"It is always a risk," he said. "Every patient tolerates chemotherapy differently and some fail it."

By the time Jane recovered enough from the side effects of Dr. T's weaponry to be discharged, she had negotiated an arrangement with Dr. S in Boulder. He agreed to treat her with the same CAF cocktail, but in fractionated doses, small enough to allow her to live as normally as possible given her now clearly terminal condition. After a port was installed under her clavicle, Jane went to his office for chemotherapy every ten days for the next thirteen months. She was able to live a moderately active life during this period. Regular doses of MS Contin (morphine in tablet form) enabled her to keep driving, shopping, and taking an occasional short stroll with Tom and Adam. Still, getting in and out of the Chevrolet van was sheer torture for her. I often found myself turning away, unable to watch.

In early May, as Addie's sixth birthday approached, Jeffy Griffin, the magical first head teacher at Friends' Primary School, showed me a self-portrait drawn by Addie. In her drawing, Addie was walking Marley on a purple leash as she moved toward a large wrapped present for her sixth birthday. Addie told Jeffy what she wanted to be inside the present: her very own Puli. In Addie's drawing, the sky was blue. But just behind Addie was an ominous black cloud, raining big drops barely missing her body. If Addie stopped moving for even a moment, she would be engulfed by a dark menacing force.

Addie Help's Self-Portrait in May 1989, walking Eben's Puli, Marley, dreaming of her sixth birthday present while being followed by dark storm cloud.

Addie's affinity for animals was rewarded by Marley's special brand of canine affection. Persistent about her Puli birthday wish, Jane finally asked me for the name of the breeder who had led us to Marley. On her sixth birthday, Addie was given her own Puli, a beautiful male, who was Marley's canine cousin; they shared a grandfather. Addie named him Duncan.

Duncan soon had the run of the Help house, impressing everyone with his superior intelligence. Something of a gourmet, cold kibble did not excite Duncan's palate. Since very few hot food items were on any Help menu, Duncan would remove kibble from his bowl near the patio doors and carefully carry pieces in his mouth across the living room to a heating vent near the fireplace, on which he placed his kibble for warming.

I smile while suppressing tears whenever I think of Duncan's creative canine cooking.

-9-

Upside and Downside of Cancer

A Bear Dream – September 10, 1989

The table is in an open field, a memory of pancakes (plain) and pancakes with something like star fruit in them. I sight seers coming down a path, looking all around them through cameras. There are pretend wooden birds hanging in nearby branches. I realize they have sighted a large brown bear and that Marley is near the bear. I call Marley and the bear comes with her. A moment of fear, but the bear rolls over and submissively offers her stomach. Then the bear rises in search of pancakes and opens a small shed door. Inside are pancakes on the ground and my pale aqua silk nightgown, too small and unworn for years.

I had followed all the orders of my surgeon and oncologist after my initial diagnosis in 1986. The threat that I would be forced to abandon my curly-haired son made me a very good patient. I weaned Eben, allowed chunks of my right breast to be cut away, and had tests and a physical exam every three months. Dr. Oliphant had referred me to Dr. R. Lee Jennings, a Denver tumor surgeon. Dr. Jennings was about my height, a comforting equality. When the pathologist reported clean margins, I was able to keep my distance from serious worry.

Dr. Jennings had referred me to Dr. T, a top Denver oncologist, who agreed to follow me along with Dr. Jennings. The low-grade nausea I that began shortly after the first two lumpectomies was treated by removing my gall bladder. The Boulder general surgeon who removed it assured me that he had carefully examined my abdominal cavity, but had seen nothing abnormal. When the nausea persisted, my internist said it would eventually disappear, suggesting it was most likely a symptom of perimenopause.

A year later, a third small tumor was found and removed from my left breast. Still perpetually nauseated, I went to see Dr. Jennings in early May 1989. A blood test showed abnormal liver enzymes; two days later an MRI showed a mass in my liver. Because Dr. Jennings was traveling, I was sent to University Hospital for a liver biopsy. My circuitous path was a reminder that cancer is crafty, calendars have conflicts and complications, and there is a vast range of diagnostic skills among doctors.

On May 15, 1989, the day after Mother's Day, Dr. Jennings called to tell me the liver biopsy performed three days earlier at University Hospital by another surgeon was positive for "metastatic mammary carcinoma." Our old-fashioned telephone receiver suddenly felt too heavy to hold, as if it weighed fifty pounds. My mouth filled with the taste of sour milk, something I subsequently came to recognize as my own fear pheromone.

After that call ended, I found myself in an emotional swamp—one toxic with fear and grief, mixed with marital guilt and shame. I felt unable to tell my husband. The mortal news came only a few days before our household was planning to travel to Florida to stage a surprise party for one hundred guests to celebrate the seventy-fifth birthday of Bob's mother, Ruth. After his mother's party, Bob would return to Boulder with Eben and Jordan, while I was scheduled to remain on the east coast for professional work.

Ruth Cantor Pelcyger with her two sons, Bob and Joel, on her 75th Birthday

I had a fellowship to attend the Annual Conference of the A.K. Rice Institute, a ten-day event at Vassar College in Poughkeepsie, New York. The conference structure was designed to examine leadership behavior and unconscious group processes in a live-laboratory, with the conference participants as the guinea pigs.[16] Following the A.K. Rice Conference, I planned to spend the weekend with Olivia before traveling to Connecticut, where I was the co-chair for a conference on legal services leadership and institutional integrity. Right after the second conference, I was scheduled to fly home to host a wedding rehearsal dinner for two of our close friends.[17]

This over-booked schedule made it easy for me to slip into a protective haze of denial—not unlike Jane. I kept my terminal diagnosis a secret through these appearances and obligations, absorbing the shock privately until I mustered enough courage to tell Bob two days after our friends took their marital vows.

After my tearful cancer confession, Bob had to leave for a previously scheduled overnight trip to Reno to meet with the Pyramid Lake Paiute Tribal Council. Early the next afternoon, leaving Eben with my father, I drove to Denver's Mid-Town Medical Center where Dr. T bombarded me with his warrior metaphors. I sat across from him, as he described his strategy.

"You have a lethal and aggressive enemy," he said looking at the pathology report. "We will use both guns on it. While it is a battle that can't be won, I will do whatever I can to cover your back while you do what you can to defend your life."

I felt frozen to the chair, even as my consciousness was staggering into a future where I imagined Eben alone and afraid.

"If the battle can't be won, how long do you think I have?" my voice already sounded disembodied.

Dr. T responded, "Since you asked, frankly in my judgment, if you are lucky and have a good response to treatment, you will survive about six months."

His words filled me with a tremulous terror. Falling obediently into line behind my commanding officer, I started chemotherapy a few minutes later. Dr. T administered CAF, his "most lethal weapon"—the cocktail of Cytoxan, Adriamycin, and FU-5.

Bob caught an early return flight from Reno, took a taxi to Dr. T's office, and drove me back to Boulder in our car. Crawling off the back seat, I staggered up our front walk—a ghost of my former self. As I reached our porch, the front door opened. In expectation of the mother he knew and depended upon, Eben greeted me with a request for immediate dinner service. Marley fixed her brown eyes on my blue ones and communicated that she too was ready for her dinner, a meal she expected to be followed by a walk.

My father, who had been caring for Eben since lunchtime, announced, "I'm late to meet the garage door repairman."

Almost simultaneously, Bob said, "I need to go to the office for a few hours."

We were all totally unprepared for the requisite role adjustments. The transition was brutal.

After my third CAF cocktail, Olivia, then in the midst of her doctoral studies at Adelphi University, flew home to help. She arrived the same day my hair fell out in massive chunks. Only then did I understand that she was as vulnerable as Eben, having mistakenly convinced myself that Olivia was old enough to be prepared for my death. I forgot that the loss of a parent is a searing experience regardless of age.

Bob and I began seeing a family therapist, Barbara B. Miller, in search of guidance for how to help our children. My memory of exactly why those sessions helped so much has faded, but Barbara's therapeutic skills were excellent. At our first session, she explained that because I, the wife, had a terminal illness, rather than Bob, the husband, our marriage was at great risk. The marital mortality rate under our circumstances was over ninety percent. (Another study showed that fifty percent of the men whose wives had a mastectomy sought divorces.)[18] These statistics indicate women are better than men at honoring the "in sickness and in health"* clause of the standard marital contract. Since all of us have an innate fear of abandonment, it was essential that Bob and I face this issue straight on with professional help.

Bob and I had an unwritten prenuptial agreement that his legal work on behalf of Native Americans, also known as "The Work Mistress,"† was his highest priority. After the first CAF treatment, I understood the full import of our implicit agreement more profoundly. I tried my best to honor Bob's deep devotion to his professional work. Other than a few post-surgical pick-ups and

---

* From the 1662 *Book of Common Prayer*.

†Over our forty-two year marriage, The Work Mistress has evolved into a powerful imaginary persona, one who almost always makes her demands at the worst possible times.

one or two key consults with physicians, I did everything I could to protect my beloved husband from the tedious vicissitudes of cancer.

Bob still has only a partial idea of what it took for me to keep functioning so that he could remain dedicated to his Native American clients, just as I have no idea how hard it has been for him to live in such close proximity to desperation and uncertainty.

Many times I came close to death by trying to demonstrate I was still a competent human being. I ignored my body's increasingly loud protests because I so wanted Eben to have a half normal childhood and to avoid adding to Bob's already too-heavy load. But, I quickly discovered, despite my fierce will, sometimes my motor simply wouldn't start. There were innumerable instances when I ran out of gas in the middle of a busy freeway of obligations, without anyone to call for help.

After lunch with friends on a day when I was too nauseated to eat, one of my lunch companions sent me an e-mail.

"I love you, my dearest. But I want my old Joan back, chocolate-eater, telling it like it is, laughing, engaging, entertaining, and brilliant. Tell me the old Joan will return, please!" she wrote.

No one missed "Old Joan" more than me. Cancer is an assault on one's identity. There is the degradation that the disease and its treatments inflict on body-image, whether it is the loss of an organ, appendage, hair, or the surgical scars. Also there is nausea, pallor, bloating, and mouth sores. The sudden changes can be shocking to ourselves, as well as to others.

Bob's standard compliment: "You look great, especially considering what you have been through!" never put my Humpty-Dumpty back together again.

My loss of status as a professional working woman was among my greatest challenges. I found it humiliating to be without my own income for the first time in my adult life while my medical expenses were gobbling up resources.

There is also the issue of social capital. When one has a large group of friends, it is easier to ignore the inevitable disappearances. The desertions of some stung even though I understood that they did not want to be reminded of our common mortality or because they felt uncertain about what to say.

There is one upside. If you develop cancer or any other major defect, you also develop new friendships with others in the same leaky life boat. A sense of connection comes quickly, like travelers who accidentally meet far from home, only to discover they have arrived from the same place. When death is close, intimacy with others changes. The distance between the well and the dying grows greater, while the distance between the dying shrinks.

The author on Eben's sixth birthday during chemotherapy, September 1989

Jane's eight children made me feel that her life was especially important to save. In my way of looking at the world, the demands of the collective dependency of Jane's progeny provided her with a biological imperative. If she died, she would leave behind a baby boy born at the moment of her diagnosis and seven others, still at home, none yet out of high school. Margaret and I would each abandon only two children. All the while I fought to keep Jane alive, I never spoke about the virtue I assigned to Jane's rank. It was just something I woke up with every morning. Her death would have greater repercussions, would be of a higher order of magnitude.

For me, it was different. I had grown up feeling permanently unwelcome on earth, the damaging side-effect of my mother's paranoid schizophrenia—always feeling as if it was necessary to make reparations, to pay rent on my existence. After I became unable to hold up my end of the tenancy agreement, let alone my wife/mother obligations, I felt that my family unit would be better off without me.

Margaret Clifford wore her anger like a dress cut too low, never covering her wounds, Jane acted as if everything was perfectly normal, as if nothing was happening. I moved into their lives because one day I had been in full-bloom as a late-life mother. The next day I was told I was going to die in six months. What was I supposed to do? How was I to prepare my children for my death? What could I do to help Bob?

As soon as I looked into Jane's life, I found I could more or less stop worrying about my own. It was also an act of desperate sublimation.[19] Initially, I didn't understand how it worked. Only later did I realize that Jane's denial was greater than my acceptance of death. She played out her denial in my hope for her recovery. Only Margaret faced reality. She knew from the first moment what was ahead and began screaming at the top of her lungs.

After our terminal diagnoses, we each had temporary reprieves. We used our reprieves differently. Jane, seemingly unable to confront her prognosis, chose to ignore all medical advice and continued in her role as a nursing mother. Margaret threw herself into a graduate degree program, dreaming of besting the therapeutic skills of her husband. My goals were mixed. I wanted to protect the devotional aspect of my husband's professional work and to develop a school for Eben that would cushion his loss and guide him for the first years of my anticipated absence.

Friends' School was where our three cancer-stricken life paths crossed, but we might have met anyway because our friends herded us together, away from them. All species do this. We clump together for comfort and protection, while excluding those who are different or defective.

-10-

In the Woods with Bambi

In early November, at the end of the fourth month of the CAF cocktail, Dr. T ordered both an abdominal CT scan and a nuclear heart study. The tumors in my liver appeared unchanged. My tumor marker had risen. Further, the ejection fraction of my heart muscle revealed what Dr. T called a serious treatment side effect known as "Adriamycin cardiomyopathy." Instead of killing cancer, the Adriamycin had damaged my heart muscle.

With the illuminated scans of my body displayed on the screen behind him, my cancer commander had "new orders" for me. He kept looking down at the notes he was writing, rather than at me.

"I am no longer able to give you the CAF cocktail. I am going to have to switch you from Adriamycin to Methotrexate."

Previously Dr. T had been adamant that Adriamycin was the most powerful weapon in his arsenal. As he wrote out the new formula for his nurse to take to his team of chemical cookers, I leaned forward to ask what I thought were reasonable questions.

"If the Adriamycin isn't working, is it worthwhile to change to a new chemotherapy drug?" I asked. "If my heart is damaged, and you say I can only expect to live a few more months anyway, why switch regimes? What reason is there to continue treatment?" my voice trailed off.

Dr. T snapped back, "I can't keep giving you a drug that will most certainly irrevocably damage your heart. If you are not going to be serious about your treatment, then I don't want to treat you!"

"What?" I said, straightening my back. "The questions I just asked were perfectly appropriate. Why can't we discuss my options rationally?"

We couldn't talk rationally because Dr. T's adherence to the doctrines of the Church of Chemotherapy had converted treatment protocols into absolutes. He would rather recruit a replacement, (and there were plenty in the waiting room ready to enlist), than face either my treatment failure or my loss of faith. Cooperative cancer patients do not confront the limits of their oncologists' arsenals. Instead, polite patients miss appointments and ignore postcard reminders.

Sensing that I was about to be court-marshaled for insubordination, I began to tremble. I meekly asked to skip the new cocktail until I could discuss the options with my husband. Without responding, Dr. T stood up, roughly pulled my scans off the screen, rounded his desk, and handed them off to his nurse, who had entered the room as if on some secret cue.

Without looking at me directly, he bellowed, "You will die a horrible death if you don't continue with my treatment!" before quickly disappearing behind the door of an exam room across the hall.

I tried to stand, but my body turned hypothermic. Dr. T's nurse brought me a cup of water and I drank. I felt as if I had been struck by a humanized weapon, and was desperate to escape the battlefield. Tears streaming down my face, his nurse finally helped me to my feet, and I staggered down the hall, past exam room doors, before crossing the wide and fear-filled waiting room. Outside I took a deep breath of cold Colorado air under a cloudless blue sky before making my way to the front of St. Joseph's Hospital, praying there would be a waiting taxi.

That night, after Eben finally fell asleep, Bob and I mutually agreed that, since I had "failed to respond to the most powerful chemotherapy," it didn't make sense to continue treatment. How could a death without chemotherapy be any worse than death in three months with chemotherapy? Given what had happened to my body and our lives over the past four months, we decided to opt for quality over quantity. It was somewhat analogous to dropping out of a hard college course with an "I" for incomplete, rather than sticking with it for the guaranteed "F." In the morning Bob called Olivia, Gale, and Jordan to ask them to come home for a blended family holiday gathering. We were certain it would be my last.

At the recommendation of Dr. Jennings, I had already met with the Hospice social worker before my first chemotherapy treatment. Within hours of my call to Dr. T's nurse informing her of my decision to end treatment, the same social worker was back at our door with the papers she needed signed for Hospice to assume responsibility—i.e. to receive insurance payments—for my care.

As Hospice staff took over, Bob and I agreed we would live as normally as possible during whatever remained of our joint future, naively believing we could prevent cancer from taking the driver's seat. We quickly discovered we had no defense against the endless stream of directives that cancer issued from the backseat. We lacked any kind of road map or GPS for our new normal, but we felt certain we were on the road to my death and it was only a short distance away. We began driving cautiously, well below the speed limit, looking carefully at every road sign, sometimes stopping to pick up fellow travelers, and pulling off at every rest stop along the way.

Bob and I had already completed the legal documents and medical directives recommended by Hospice. All that remained were cremation arrangements, as well as the letters the social worker suggested I write to Bob, Olivia, and Eben, as well as Gale and Jordan. A Hospice nurse arrived the next day to examine me, chart my vitals, and arrange for comfort medications—anti-nausea, pain, sleep, and the anti-depressant Zoloft, an antidote that Dr. T had prescribed when I started chemotherapy without my realizing its purpose.

Bob and I both felt as if we had been sealed inside an impenetrable bubble—one that allowed us to see the world as we had known it, but that kept us separate. Being inside the bubble obliterated any sense of the future, except what we had to do in the next few hours or what was required to make it through the night. And now the first fork on the treatment trail had come to an abrupt dead end in a deep dark forest.

Friends' School was closed the day after Thanksgiving. With totally inadequate parental forethought, motivated mostly by my hope for a mid-day nap in the darkened theater, I took Eben to a matinee screening of Walt Disney's *Bambi*.

We stopped for groceries after the film. During the ride home and while shopping, Eben was unusually silent. Bob was home when I parked in front of our house and Marley alerted him to our return. Eben quickly freed himself from his car seat, jumped out, and ran up our front walk, just as Bob and Marley appeared on the porch.

In a voice that sounded as stern and loud as that of Dr. T, Eben shouted, "Dad! The mom dies and you have to take care of me!"

Eben remained obsessed with Bambi for many months. After he found a pair of velour antlers attached to a plastic headband in the Friends' School dress-up trunk, he insisted on wearing them day and night. He adaptively played the role of "Bambi in the Manger" in the ritual re-enactment at the Friends' School Christmas celebration. During recesses, block play, and whenever dress-up clothes were pulled out, Eben relentlessly tried to recruit his playmates to take up the roles of the father, the mother, and, most importantly, the hunter. Eben also asked that *Bambi, A Life in the Woods*[20] be read to him as often as three times a day. It was agonizing to see my beloved son try to come to terms with what he intuitively knew was approaching.

Eben with Bambi totem given to him by his teacher Jeffy Griffin, Friends' Primary School, Boulder, June 1990

Our close ONINer* friends guffawed when Bob surprised me with an unusual holiday gift of two burial plots in Boulder's historic Columbia Pioneer Cemetery, located just two blocks south of our home. They gleefully declared it to be more gauche than the gift of a yogurt maker that David H. Getches presented to his wife, Ann, on her thirtieth birthday. While the body of another gentleman was resting between the two plots, Bob and I were not concerned

* ONINer stands for Original Non-Indian NARFers—a group of five couples, of whom at least one partner had worked at NARF during its formative years.

about the distance between us after death. Instead, we were concerned that Eben might need a place nearby to visit for several years as he adjusted to my absence.

In December, we sat for what we thought would be our last photograph as a blended family. Everyone had their picture taken while wearing my wig, which was so tight on me that I couldn't tolerate it for more than a few minutes. We unanimously agreed that my father looked best in it. Then we put on alligator noses to hide the stark awkwardness produced by the fear of death, colored by feigned normalcy.

Family gathering after author became a Hospice patient, December 1989. L to R: Gale, Jordan, Bob holding Marley, the author, Eben and Olivia.

-11-

The Tree of Life

The holidays ended, my mouth sores partially healed, and I began to feel less fatigued. Surprised to be still alive in January, I began ingesting a poisonous potion called extract of *Thuja* from *Arborvitae,* or the Tree of Life. It reminded me of an earlier medical experiment in Berkeley with another poisonous plant, the abortifacient mistletoe.

I wasn't alone—twelve other terminal patients in Boulder joined me in one of those spontaneous, non-FDA approved experiments which draw desperate cancer patients. They usually occur in association with some kind of spontaneous remission, after which a kind of fairytale develops about the magical properties of one thing or another. Our desperation to survive, mixed with limited treatment options, leads us into the wilderness of amateur science.

The Tree of Life experiment had its origins in a Boulder breast cancer support group that Jane, Margaret, and I had joined. At those meetings, I had met and become close with another member, Nuhiela Audeh. Born in Palestine, Nuhiela had come to America as a new bride, after her husband, Azmi, was admitted to graduate school in Kentucky. The couple moved to Boulder when Azmi was hired as an engineer for a local computer company. They had two almost grown daughters and had just begun to enjoy retirement when Nuhiela was diagnosed with breast cancer.

During one meeting, Nuhiela told the group she was certain her breast cancer had spread, even though she had bilateral mastectomies and chemotherapy in an effort to prevent a recurrence. Still, Nuhiela's anxiety was not responding to the assurances of her doctors that she was cancer free and would be fine. Nuhiela's angst was not uncommon—cancer frequently produces a secondary case of hypochondria.

As it turned out, Nuhiela's fear was reality-based. It was not long before her Denver oncologist confirmed that her breast cancer had metastasized. Desperate to survive, Nuhiela went to see her breast surgeon, Dr. John Day, at the Boulder Medical Center because he was rumored to have an interest in alternative treatments. Dr. Day thought of another of his patients, Thijs DeHaas, a scientist at the National Bureau of Standards, a federal laboratory in Boulder. A few years prior to performing Nuhiela's mastectomies, Dr. Day had removed a Stage III melanoma growth from Thijs, after which he referred Thijs to Dr. Bill Robinson, the top melanoma specialist at the CU School of Medicine. Dr. Robinson had made an error during Thijs's initial consultation. Looking at Thijs's wife, Dieuwke, who was Snowball-like, with white blond hair, Dr. Robinson declared her to be a "perfect specimen for melanoma." Deeply offended, the couple left immediately and never returned.

Instead, Thijs DeHaas traveled to Holland, where his uncle, a physician and a pharmacist, had developed a remedy to treat several cancer patients, apparently successfully. Thijs acquired several dozen bottles of his uncle's remedy, as well as detailed written instructions for its formulation. It was an extract of *Thuja,* a species more commonly known as *Arborvitae,* or the Tree of Life.

Each time Thijs came for a check-up and remained cancer-free, Dr. Day paid more and more attention. He told Nuhiela that, as far as he knew, Thijs's melanoma was still in remission. Thijs very generously gave Nuhiela some of his limited Thuja supply in June 1989. Six months later an MRI showed no signs of metastatic disease in Nuhiela's liver.

Thijs DeHaas, Hanukkah party, December 1989

I was still being treated with chemotherapy by Dr. T when Nuhiela called me with her miraculous news. That night I drove to the home of Thijs and Dieuwke in search of some of the *Thuja* potion for Jane, whose survival I believed should be at the top of any list. Thijs and Dieuwke listened respectfully to my request and gave me several bottles of the *Thuja* formulation. Although late to be dropping by, I drove directly to the Help home. After telling Jane and Tom about Nuhiela's remission, Jane took her first dose of *Thuja* in my presence.

I had stumbled into my initial meeting with Thijs DeHaas without knowing anything about the formulation of his character. He was a man of tremendous compassion, generosity, and energy. Within a week of our first meeting, of his own volition, and at his own expense, Thijs started driving to the East Coast to harvest branches of *Arborvitae* from trees at a certain latitude to ensure maximum levels of poison would be present in the needles.

While Thijs was away, Dr. Day and I began searching for a way to produce the formula to the appropriate standard in Boulder. Professor Larry Gold, the father of Nicholas, Eben's best friend, kindly persuaded Dean Stull, the founder of Hauser Chemicals, to take on the manufacturing task. Larry also persuaded his good friend, Charley Butcher, owner of the Butcher Wax Company, to pay Hauser ten thousand dollars for production costs.

Thijs returned from Connecticut with a dozen large gunny sacks full of *Arborvitae* cuttings and delivered his harvest to our home. For a full week, my father and I sat in our living room, carefully removing the budding tips of the *Arborvitae*. Eben assisted, as did Bob at night. For years afterwards, Eben was preoccupied with concocting potions in his play.

As soon as we delivered the *Arborvitae* tips to Hauser Chemicals, Dr. Day and I began searching for other Stage IV cancer patients to join in a small clinical trial. By mid-January 1990, we had recruited a dozen terminally ill patients, including some with well-established academic reputations. I wanted our efforts to at least meet minimal scientific standards. I also felt that other well-educated patients provided some modicum of respectability.

The purpose of the trial was to see if we could replicate the responses of Thijs and Nuhiela. Dr. S and his partner met with Dr. Day and me, but their reaction was negative. They finally agreed to allow us to leave information about the *Thuja* trial in their waiting room. We made a presentation to those interested in the experiment in a conference room at the Boulder Medical Center. Dr. Pat

Moran, then an oncology resident at CU's University Hospital, was the only oncologist who attended. I did not contact Dr. T.

In addition to Nuhiela, Jane, Margaret, and myself, the trial included Walter Orr Roberts, Professor of Astronomy and Atmospheric Physics, who helped found the National Center for Atmospheric Research (NCAR); Stan Gill, Professor of Organic Chemistry; and Naomi Grothjan, Principal of University Hill Elementary, and five other patients. Over forty patients had applied. Two tablespoons of the *Thuja* potion, formulated in an alcohol-based liquid the consistency of cough syrup, were taken every four hours. Each trial participant wore a wrist watch with pre-set alarms to assure accurate compliance and consistent dosing.

Less than a month went by before the first patient died, a thirty-five-year old woman with lung cancer, who smoked on her last the day of life. Her death was not surprising since, just as in most initial drug trials, all of the *Thuja* study participants had to have been diagnosed with Stage IV disease, indicating a short life expectancy. When this first death occurred, we all had a new understanding of why clinical trials are blind; our magical thinking was disturbed.

A few weeks later, on March 12, 1990 death knocked again; Walter Orr Roberts died. Soon thereafter Naomi Grothjan left Boulder for Boston to undertake treatment at an Ayurveda clinic. Naomi represented the first and only defection from our patient-developed *Thuja* trial. She died before the end of March.

Other deaths followed, but most of the patients lived at least six months longer than predicted and more than half showed measurable shrinkage in their tumors. All but three of the patients in the trial outlived the prognoses of their doctors, including Nuhiela, Jane, Margaret, and me.

I am now the last survivor of the *Thuja* trial. Thijs DeHaas died of Alzheimer's disease in August 2005. Many oncologists and statisticians would argue that two survivors out of twelve is not a bad statistical outcome for a Stage I clinical trial, even if the formulation was inert and responses due only to the placebo effect. Regardless, the drug Taxol and its derivatives are now used to treat many cancers. Taxol was initially manufactured by Hauser for commercial use using cuttings from *Thuja*. Today synthetic copies of *Thuja* molecules are used, as well as for newer formulations, like Taxotere and Paclitaxel.

-12-

Alternative Treatment: Metabolic Therapy

A few months into the *Thuja* trial, Hope Morrissett introduced me to Ginny Wells Jordan, a Boulder Jungian dream therapist and the mother of three young children. Ginny had been treated for breast cancer[21] with both surgery and chemotherapy. While Ginny did not have metastatic disease, her parents had hired a medical consultant to search for adjuvant treatments hoping to prevent a recurrence. The consultant had recommended a metabolic program directed by Dr. Nicholas Gonzalez in New York City. Ginny encouraged me to try the program.

"Joan," she said, "It requires no chemotherapy, just healthy stuff, including gallons of carrot juice, vitamins, and raw grains!"

Seeing skepticism in my eyes, Ginny provided me with the names of two other metastatic breast cancer patients being treated by Dr. Gonzalez. After long distance telephone conversations with those two patients, Bob and I decided to risk the fifteen hundred dollar fee for the initial consultation.[22]

The office of Dr. Nicholas Gonzalez, a graduate of Brown University and Cornell Medical School, was in a street level suite on Park Avenue and East Seventy-Ninth Street, marked by the traditional prestigious brass plate. The location and the brass plate were New York City status symbols providing reassurance for his anxious and desperate patients. At that time, Dr. Gonzalez was accepting "heavily pre-treated" patients like myself out of economic necessity. I entered his office skeptically because, while I was no longer active in the Church of Chemotherapy, my faith remained in orthodox cancer treatments.

At the end of the first half of my two-day Gonzalez consultation, I called Bob and told him the treatment was too far from my belief system to proceed, so I was going to come home in the morning. A few hours earlier, while sitting across from Dr. Gonzalez, I felt like I was Dorothy meeting the Wizard of Oz. My father had instilled in me a strong instinct for what is plausible, and I found it hard to believe that there was anything more than remedies my Mormon pioneer ancestors might have used behind Dr. Gonzalez's curious curtain.

Bob pleaded, "Joan, fifteen hundred dollars is nothing compared to what it might mean if he can help you!"

Bob insisted I call the daughter-in-law of a close friend of his mother, who was also being treated by Dr. Gonzalez.[23] After she provided explanations for some of my most troubling impressions, I decided to stay for the second half of the consultation.

The next day I listened very intently to what Dr. Gonzalez had to say. Looking back, it is possible Dr. Gonzalez had reached something separate and secret from my rational mind.

I remember well the moment Dr. Gonzalez said, "Ms. Lieberman, you can get better if you want to!"

Dr. Gonzalez spoke as if his words could cut through the thick noose around my neck. Despite my cynicism, something in his persona, his authoritative surroundings, and his rituals may have spoken directly to my unconscious, triggering some kind of cellular commotion.

Back in Boulder, I woke up every morning surprised to still be alive as I struggled to adapt to Dr. Gonzalez's "General Dietary Guidelines for a Moderate Vegetarian Metabolizer"*—one of several dietary categories into which his patients were divided. I ate mostly raw vegetables and fruits, had a daily breakfast bowl of fourteen raw grains soaked in organic apple juice, washed everything down with carrot juice, and hastened the exit of all the by-products with at least three coffee enemas a day. In between, I swallowed hundreds of enzymes, vitamin pills, and mineral supplements, including ground up pork pancreatic "enzymes" which I gulped down four capsules at a time, six times each day between meals and in the middle of the night.

Compared to chemotherapy, Bob and I found Dr. Gonzalez's treatment methods both demanding and comforting. There were pills of all colors, shapes, and sizes to sort into miniature Zip-lock bags; vegetables and fruits to juice; and

---

* Dr. Gonzalez divided his patients into genotypes. For example, an Eskimo patient would not be treated with great quantities of carrot juice, but rather would be supported with a native diet, including whale blubber.

grains to grind and soak. I was also required to use good coffee for less-than-sacred purposes.

It was simple and staggering, comforting and complex, nasty and normalizing—all at the same time. There were so many rules and rituals, it was like becoming an Orthodox Jew. Or, if one were an ex-communicated Mormon, it resembled religious regression to a much more extreme version of the Word of Wisdom,[24] a rule, prohibiting alcohol, caffeine, and allowing for only sparse consumption of flesh.

For all practical purposes, Dr. Gonzalez's protocols required me to become my own doctor and nurse. One major advantage was that treatment was at home, as opposed to hospitals or doctor offices, so I could continue my mothering of Eben. Bob, having never been an active member of the Church of Chemotherapy, was an enthusiastic convert to Dr. Gonzalez's Church of Carrots, Coffee, and Pancreatic Enzymes. He even energetically juiced carrots for about a week. I had never eaten foods as healthy as those prescribed by Dr. Gonzalez. While I am quite certain that they didn't cause me any harm, I remain ambivalent about whether any contributed to my survival.

Nonetheless, I was a dutiful soldier in Dr. Gonzalez's army of desperate volunteers and chemo-deserters. I tried hard to do everything Dr. Gonzalez ordered, but I believe total compliance is impossible. Considerable time was spent lying low, literally, on the cold slate tiles of our bathroom floor, ruining my relationship with coffee.

One day while I was prone on the bathroom floor, I overheard Eben's friend Nicholas Gold ask, "Where is your Mom?"

Eben replied, "Oh, she is upstairs doing her enemies."

"Who are her enemies?" asked Nicholas.

Eben's retort, "Oh never mind! They aren't that interesting!"

While I used a bath towel to cover myself, Eben would lay down next to me so I could I read to him with periodic private intermissions. Whenever I asked him to leave the room, he did so promptly—as if he were a Help child.

-13-

Adjuvant Therapy: Hypnosis

When I began chemotherapy in 1989, the remedies for nausea were problematic compared to the drugs that have since been developed. The anti-emetics at that time reduced nausea by making me so sleepy I was dangerous as a mother, driver, and cook.

In an effort to care for Eben, as well as to honor The Work Mistress clause in my marital contract with Bob, I tried acupuncture with Warren E. Bellows. He was known to have treated Treya Killam Wilber, a well-known breast cancer patient in Boulder, who had rated Bellows highly. No doubt Bellows' skills were excellent, but after the second session in Boulder, he asked me to travel to Denver every day for treatment at his new office. He might just as well have asked me to cross the Atlantic in a small rowboat.

After acupuncture became an impractical anti-emetic, Barbara Miller, our family therapist, suggested, "Maybe you should see Roland Evans. He is an Irish psychotherapist, who works with Hospice patients and uses hypnosis to aid in symptom control."

As Barbara spoke, I silently scoffed. But desperation is an extremely effective antidote for skepticism and resistance. Putting myself under the care of a hypnotic Irishman was mostly about Bob and me being able to keep at least our toes in the shrinking puddle of denial. I made an appointment.

It was a challenge to climb the stairs to Roland Evans' office, then located on the second floor of a building directly east of Howe Mortuary—the site of my soliloquy to my mother's remains. When I reached the top of the stairs, I decided Roland Evans must be seeing a very hardy type of Hospice patient in a building that could best be described as "hostile to the handicapped." The communally decorated waiting room was filled with a diverse collection of cast-off chairs. Before I chose one, I went into the bathroom to vomit. Then with all the dignity I could muster, I sat down to wait.

From the sound of Roland Evans' accent on the telephone, I was expecting someone who resembled Alistair Cooke, the British broadcaster—someone older and wiser than me, a benevolent father figure. Actually, a biblical version of God in modern clothes would have been good.

Instead, after an appropriate interval, a young man who looked like he might be an exchange student at Boulder High School appeared. He had a soft fuzzy quality about him, as if he were a new born animal. His hair, eyes, and sweater all spoke of cuddly innocence.

I remember thinking, "Oh No! He is young enough to be my son! No way is he going to be able to help me! He can't possibly have had enough life experience."

I wanted to excuse myself, even as I reached out to shake his hand. In my head, I was running away, offering to pay his fee as I backed out of the waiting room. Yet, I believe I am alive today because something in the eyes of Roland Evans would not let me give up so easily.

Bald and bloated, I had been forced to dip back into my maternity wardrobe because of ascites* from my malfunctioning liver. A tender scalp had made me adverse to my too-tight wig, so I had resorted to wearing a soft scarf secured by an artificial braid on top of the scarf. These wardrobe adaptations caused me to be mistaken for a pregnant Muslim woman on more than one occasion.

Fortunately, Roland Evans did not pay much attention to my appearance, and I was mistaken about both his experience and age (he is only nine years younger than me). Chemotherapy not only makes you feel older, it makes you look older. But by the grace of the universe, I also did not know that Roland Evans would turn out to be as comforting as my first best friend, Marlene Evans. Perhaps the Evans effect on me is genetic. The ancestors of both are Welsh.

Sitting across from Roland Evans, whose face is both handsome and perpetually kind, I was convinced I would be immune to hypnosis, that the technique was a waste of time and money. Instead, I discovered it was extremely effective because I was super susceptible. I was stunned when the sound of Roland's voice produced an unexpected sensation of pain-free coolness that started in my feet, moved up my legs, through my abdomen and settled in my liver. In the beginning, I saw Roland several times a week, then sometimes more.

---

*Ascites is a medical term indicating the accumulation of fluid in the abdomen.

Later Roland made an audio tape for me to listen to on my Walkman to provide relief wherever and whenever needed.

Not long after I began seeing Roland I came to think of myself as the equivalent of a hypnotic slut.

Roland Evans eventually became an instrumental appendage to our entire family, a human talisman providing comfort and understanding. He provided therapeutic support to both Eben and Bob during their times of need and he remains an invaluable asset and witness to our highly improbable family story.

I am certain I would not be alive had I not discovered my hidden susceptibility to hypnosis under Roland's expert and humane tutelage. I owe him an enormous debt of gratitude for his remarkable therapeutic skills.

Roland Evans

-14-

Secret Spending

Soon after the beginning of the second year of Friends' Primary School, Jeffy Griffin, the Head Teacher, called to tell me her husband had seen a notice of foreclosure on the Help home in the *Boulder Daily Camera*. Still a novitiate in Dr. Gonzalez's treatment monastery, I felt steady enough from some mysterious combination of *Thuja*, Dr. Gonzalez's protocols, and hypnosis to drive across town to the Help house without calling.

Jane was sitting on one of the chenille-covered couches with a pile of bills in her lap. The handset of the princess wall phone, cord stretched taunt, was in her hand. I have a direct communication style that occasionally gets me in trouble. A close friend once described me as "prone to blurting."

Getting right to the point, I asked, "Jane, what is happening? Is it true your house is in foreclosure?"

Jane confirmed that foreclosure proceedings had begun, but she was more worried about a final warning notice from the Public Service Company that the gas and electricity were going to be turned off before dark. I asked to see the Public Service notice. As Jane struggled to give me the document, I realized the first item of business was to get her to the hospital; Jane not only was in obvious agony, she had lost control of her bladder and bowel.

After calling an ambulance and Dr. S's office, I reached Tom at his place of business to tell him what was happening and to make certain he could meet Jane's ambulance at the hospital.

Tom's response, "How long do you think I will be there?"

Denial was his default mode, although cancer is nothing if not inconvenient.

As the ambulance siren faded away, I began gathering legal documents and bills. Looking for something to hold them, I caught site of Jane's quilting bag. Inside I discovered a horrendous collection of unpaid credit card bills, pharmacy receipts, and confusing insurance statements. To postpone facing the full scope of the problem, I emptied the dishwasher, fed Duncan warm kibble, and took out the garbage. I also left a telephone message for Jane's mother, Jeanne, who had moved to Boulder from California to help care for her daughter, but was having a rare day off.

With the bills in the quilting bag, I drove to the Public Service Building on Canyon Boulevard and paid the arrears, plus a deposit. The clerk promised me the power would be on when the children came home from school and before the meager contents of the Help freezer and refrigerator were spoiled.

The next twenty-four hours were dedicated to finding a way to prevent the eviction of the Help family and the loss of their home. In the end, having given up on my own future, I emptied my Individual Retirement Account (IRA) to cover the four-month arrears that had brought the Helps to the brink of eviction.

This was my entry into a more subterranean aspect of the Help family. Not only was Tom ignorant of the foreclosure threat, Jane did not want to tell him. Many women have trouble telling the truth about money, particularly what it really costs to keep a household afloat. Add terminal cancer and eight kids to this gender-based vulnerability, and the risk of deceit increases. Certain that I would never reach retirement age, it was easy for me to sacrifice my IRA. At the specter of Jane's eight children becoming homeless, I didn't hesitate for a second to reduce the size of my own children's estate. Nor did I tell Bob what I had done.

At the hospital, an MRI revealed that the tumor once again had its clutches around Jane's spinal cord. After being admitted, Jane was introduced to two surgeons—one neuro and one vascular—who stood at her ICU bedside to explain how they would try to prevent further paralysis. Their plan was to open her body from both the front and the back to increase access to the tumor, hoping to avoid further damage to the cord. Jane told them she needed a day's delay so she could see her children before the surgery. The two surgeons argued that every hour of delay increased the risks of irreversible damage. The patient prevailed.

When I brought Addie and Katherine to the hospital after school, Jane was holding a scrap of paper on which the telephone number of the legal firm representing the mortgage company was written. Tom, who had been at her side most of those hours, never asked her about this clue. When Tom went down the hall to call his place of business, I sent the girls and Eben across the street for a bagel. The lawyer for the mortgage was demanding payment of all court costs in addition to loan pre-payments for the next two-months before he would agree to withdraw the foreclosure filings. With Eben, Katherine, and Addie in tow, I drove to the bank to wire the additional amount.

Many might judge my role in bailing out the Help family as false heroism or fiscal stupidity, maybe both. But there would be very little heroism in the world, neither high, nor low, neither false, nor true, if every human being gave due consideration to all risks before stepping forward. I see now that I was still spending what was left of my life trying to save another mother—not my own, not one in Africa, but one who was vitality important to eight children.

Jane survived the ten-hour surgery and two weeks in the ICU. The worst moment came two days after the surgery. Tom and I were at Jane's bedside when the neurosurgeon pulled back the curtain of her ICU cubicle. Standing as far from her as possible and looking past her at the wall, the neurosurgeon said that although he and his colleague had removed as much of the tumor as possible, they had been unable to remove all of it. Still avoiding making eye-contact, the neurosurgeon said he had just spoken with the pathologist and with Dr. S.

"I am sorry to tell you, Mrs. Help, but we are in agreement that it is unlikely you will live more than three months." Then, without pausing, the neurosurgeon quickly moved on to a more technical description of the surgical results, "Jane, you are now a T-4 paraplegic. As soon as you are able, we will transfer you to the Mapleton Center for six months of rehabilitation and training."

This level of contradiction in diagnostic feedback is not uncommon, but this particular prognostic narrative was off the chart on the crazy-making scale.

While Jane was still hospitalized, I arranged for other mothers to substitute for me so I could keep an appointment made months earlier to attend a mini thirtieth high school reunion with three of the four friends with whom I had traveled to Europe in 1961.* The fourth classmate had generously rented a condo for a weekend reunion at Angel Fire, New Mexico, instead of the so-called real reunion in Bakersfield. During my eight-hour solo drive south through Colorado's beautiful Middle Park, I found myself ruminating about women and secret spending, my own included.

---

* Nonny Thomas, Kim Brown Federici, and Susan Briggs. Sherry Woods died during surgery in 1970.

That night after dinner, I posed the question about secret spending to the reunion group. All of my high school friends had similar stories to tell. We were still talking about the phenomena the next morning when we went into the closest grocery store. Standing in the checkout line, another female grocery shopper overheard our discussion and spoke up.

"What I do is keep several old dry cleaner bags in the trunk of my car. Whenever I buy a new outfit, I discard the store packaging and put a dry cleaning bag over my new clothes. That way, when I bring it into the house, if my husband sees me and asks, 'Is that a new dress?' I say, No, I just picked it up from the dry cleaners."*

Jane did go to the Mapleton Center for rehabilitation, but only for five weeks. She asked to be discharged before Christmas to be with her children, a need no one contested. Tom hurriedly built a ramp for the front steps and the necessary equipment for her care was ordered.

An ambulance brought Jane home and the attendants gently lifted her onto the hospital bed, now center stage in the Help living room. Jane used the overhead hoist to pull herself into a sitting position and Adam quickly joined her on the bed. It was three o'clock in the afternoon. With her youngest child beside her, Jane put her discomfort aside to listen to the school reports of her older children.

For the next year, with considerable help from her mother and friends, Jane's will to survive served as a balm for her children.

---

* In an ironic coincidence, a year later the fourth classmate was arrested and convicted for having embezzled funds from her employer. She had used some of those funds for the rental of the condo in Angel Fire.

-15-

Two Different Visions

In January 1991, the Board of Directors of Friends' School asked me to write a grant proposal for the expansion of the school. America had just gone to War in the Persian Gulf in response to Saddam Hussein's invasion of Kuwait. I felt a need, in the midst of both my own and the world's uncertainty, to fashion something, to drop it into the confusion—to make an offering to the life force.

Despite having written hundreds of proposals, I was inexplicably stumped about how to describe an institution conceived in my heart and carried in the womb of my home. Then, in the middle of February, I awakened from an afternoon nap filled with dream images of Friends' School in an imaginary future. I started writing, describing in detail my hopes for how the School would develop over the next five years, finding it hard to stop. Fifteen days later, on March 1, 1991, just after one o'clock in the morning, I finished a book-length proposal entitled, *A Vision for The Future—Imagining a Visit to Friends' School, A Children's Center for Learning the Ecology of Living.*[25]

Anxious to complete my now long-overdue Board assignment, I took the manuscript across town to Kinko's, a twenty-four-hour print shop. I drove home and crawled into bed next to Bob. I fell asleep easily, but a few hours later I was awakened by what I could only describe as a "Being of Light"—a luminous androgynous figure standing over seven feet tall. Paralyzed with fear, I was unable to move or speak. When the figure moved closer to me, my body felt as if it was being electrified. Then, as the hands of the Being of Light passed above my body, I felt a sudden sense of pain-free well-being. Only after the figure turned and began moving away from our bed, did my paralysis pass.

Filled with excitement, I shook Bob awake and tried to get him to focus. "Bob, do you see it?" I pleaded. "Look, right over there! Oh, no! It's going away!"

Then I began begging the Being of Light, "Whoever you are, please don't go yet!"

Bob, who has a well-deserved reputation for waking with less than full mental acuity, finally roused himself enough to understand what I was saying. He said he saw nothing, but to me our bedroom still seemed full of light. I remained wide awake the rest of the night.

As soon as the sun came up, I left the house to pick up the copies of the proposal. Waiting at a long traffic light, I wondered if I had been healed. But by early afternoon, I had convinced myself that the Being of Light experience was evidence of a psychotic break. Waiting outside Roland's new office, I feared I had become my mother or even worse—a mutated version of Joseph Smith imagining a visit from the Angel Moroni.

Roland was helpful, assuring me that many people reported similar experiences. Further, my current thinking patterns and behaviors were inconsistent with psychosis.

"You are not crazy," he assured me. "What you experienced may be part of a healing process and a reminder that you need to explore your own spirituality."

During our very first session, Roland had asked me, "Do you have a spiritual advisor?"

I had assured him that I most definitely did not.

Slowly, over the next several days, the intense feelings generated by the Being of Light dimmed. Roland's techniques, combined with Zofran, a new antiemetic drug, made me feel much better.

-16-

A Secret Detour on the Treatment Trail

Ten days later, Paul Helmke, then the Mayor of Fort Wayne, Indiana, called to ask if I would consider working with a committee of three hundred Fort Wayne citizens to help them create a new vision for the city.[26] Most consultants consider assignments undertaken on behalf of a committee, let alone a committee of three hundred, to be professional death traps. But because I was drawn to the synchronicity of two different visions—a vision for Friends' School and a vision for Fort Wayne—I agreed to Paul Helmke's request.

Bob frantically tried to talk me out of resuming work, let alone work that was as far away as Fort Wayne, pleading, "Joan, please stay home and do your enemas!"

"Sweetheart, you have made much better arguments!" I responded.

I traveled light to Fort Wayne. The initial interview required only an overnight in Fort Wayne. Nonetheless, out of commuting shape, I was wilted when I checked into the Fort Wayne Hilton. In the morning, I was the last of three consultant candidates to be interviewed by a subcommittee of business, government, and labor leaders. As soon as the selection committee gave me a heads up, I wondered whether Bob had been right. Yet it felt wonderful and stimulating to be back on my horse, sitting in my professional saddle again.

The Fort Wayne consulting assignment pushed me off the Gonzalez program. One cannot live a normal life and also be a Gonzalez patient. After I returned to Boulder on March 18, 1991, I had the following dream:

The Mandala

I was being called to account for damage I had done to a woman named Joan. A team of doctors was called in to give their pronouncements on the extent of the damage after having examined the woman for a year. Several judges were also present who were to decide my penalties based on what the doctors said. All of the doctors and all of the judges were men.

> *The woman I had damaged also was present. She was tall and dark. The doctors announced that they could find no damage at this time and that we would have to wait another year. In the meantime, the judges instructed me to begin drawing a mandala.*

Relating this dream to Roland, I said, "I don't remember what a mandala is. It seems like it is some kind of Tibetan thing. Do you know?"

Roland smiled, maybe at my ignorance. "It's an age-old symbol for a portrait of the self. Haven't you read Jung?" he asked. "He has written about and described mandalas in much detail."

I had read Jung, but, whether it was chemo-brain or something else, I could barely remember the themes in Carl G. Jung's *Memories, Dreams, and Reflections*. Nonetheless, I went home with instructions from the dream judges and Roland to draw a mandala. I found the task extraordinarily difficult. After drawing a circle, I struggled to put any kind of image inside it. The contents of the circle seemed sparse and lacked symmetry. I drew a small cottage with a garden and a deer in front, behind it a range of steep mountains covered in pine forest. Near the forest there was a bear at the edge of a pond filled with fish—in the sky a flock of birds.

After struggling to create my mandala, I had a series of disturbing dreams about bears, Raggedy Ann dolls, sticks, dogs, excrement, rickrack, and other mysterious images. I often awoke drenched from these dreams. This period was followed by a strange dream hiatus. While I found myself longing for a dream connection, I was also feeling happy and competent—it was great to be out in the world again in a place where no one knew I was dying.

Between trips to Fort Wayne, I began planning an August party to celebrate my father's eightieth birthday. Because the Fort Wayne project was very demanding, I soon found myself avoiding most of Dr. Gonzalez's more rigorous protocols, easier to do when I was away from Bob's scolding glances in his self-appointed role as a Gonzalez Protocol Enforcement Officer.

In late June, during my fourth visit to Fort Wayne, I woke up with a dark spot in my left eye. It was the day after the first full meeting of the client group, or what was had become "The Greater Fort Wayne Consensus Committee." The meeting had been designed as a "search conference" in an effort to make it possible for all of the three hundred Committee members to actively participate.[27]

I started home on Thursday, terrified of going blind. Between planes at O'Hare in Chicago, in a double breach of both Hospice rules and alternative treatment doctrines, I called Dr. Jennings, instead of Dr. Gonzalez or Hospice. Dr. Jennings arranged for an emergency MRI of my brain and I took a taxi directly from Stapleton Airport to St. Luke's Hospital. He also gave me the name of his first choice for a neurosurgeon.

Because Bob was out of town, the next afternoon my father drove me to see the neurosurgeon, who recommended surgery with a CyberKnife, a then new kind of radiological laser beam that zapped tumors in hard-to-reach places.

After Dr. Gonzalez received the radiologist's report, he called, asking for a confession of the ways in which I had failed to follow his treatment.

"Get back on the full protocol," Dr. Gonzalez said, "and you will do fine, but I am going to make some adjustments in your program."

Once again, Bob and I found ourselves faced with having to choose between treatment ideologies. I was still a Hospice patient, but if I was going to pursue any kind of non-palliative treatment, I would have to withdraw from Hospice. Many AIDS patients had begun to do so as a result of the development of protease inhibitors and other anti-viral medicines.

Withdrawing from Hospice required approval from Prudential, the carrier then providing insurance to Bob's law firm. Prudential determined it was not obligated to provide care beyond those services offered through Hospice. The cost of the brain scan and consult with the neurosurgeon were ours alone. Given everything, including the risks, utility, and cost of the CyberKnife treatment, Bob and I decided to forego the latter and to continue with Dr. Gonzalez.

The first symptoms the lesions in my brain were progressing came a week after I called Dr. Jennings from Chicago. I lost all but peripheral vision in my left eye and suffered my first grand mal seizure. I was given Dilantin and Decadron in an effort to control the seizures. These two drugs, combined with MS Contin, made me feel as if my brain was encased in Jell-O. They did more than impair my cognitive abilities; they totally impaired my dream life. My dreams were what had given me hope that something larger was at work.

I had a professional obligation to forfeit the Fort Wayne work. Fortunately, I had already brought in Jim Krantz, a colleague from the Center for Applied Research at the Wharton School of Business in Philadelphia, to assist me with staging the search conference. Jim kindly agreed to assume responsibility for the Fort Wayne contract.

I was no longer able to drive. Hospice was suggesting I might want to check into their facility to avoid traumatizing Eben. We had to face the need for a support structure based on a coma period followed by my imminent death.

Friends' School had outgrown the little house next door in 1990. With the incredibly generous financial support of Larry Gold and Hope Morrissett, a much larger building had been purchased six miles east of our home on the opposite side of Boulder for Friends' Primary classes.

We patched together Eben's transportation for the remainder of the 1991 school year—a combination of Bob, my father, Hope, Bev, and an occasional taxi. But something simpler that could be managed by Bob as a single parent was needed for the long-haul. We reluctantly made the difficult decision that Eben would have to change schools, despite the fact that he was the original inspiration for its creation. We decided that enrolling Eben at Mapleton Elementary School was our best option because of its location. After my death, Eben could walk to school and back from Bob's office at the Canyon Center when Bob was in town. When Bob was out of town he could walk from to school and back from BaBa's condominium. This decision was understandably extremely unpopular with my co-founders and utterly heart-breaking for me. Some of the wounds from our decision to withdraw never healed.

After my first grand mal seizure, I wore a Medical Alert bracelet in the hope that no uncovered or extraordinary measures would be taken. A second seizure came mid-hair trim in a beauty salon, where beauticians automatically called an ambulance. Both the ambulance attendants and ER personnel at Boulder Community Hospital engaged in a number of expensive procedures.

Despite the assurances of Dr. Gonzalez that even brain tumors responded to his treatment protocols, my faith faltered. Deeply depressed, I believed Bob would be better off if I let go of life. He was in a constant state of worry and over-work, and our debt load was growing faster than an aggressive cancer.[28] While I knew Bob would miss me, I also felt confident that he would find a new life partner.

I had a third grand mal seizure at home while Bob was away. My father happened to be in the kitchen with me when the seizure began. He panicked and dialed 911. Dr. Gonzalez's approach to treating the brain metastasis, among other adjustments, had been to increase the number of proscribed daily enemas to twelve. But one side effect of a dozen enemas is dehydration, making it extremely difficult to maintain a therapeutic level of Dilantin, thus increasing the risk of seizure.

After being re-hydrated by the paramedics and becoming more cogent, my father began sobbing, saying that he couldn't stand it anymore. He thought I had to try the CyberKnife. Tears streaming down his face, he begged me, promising to cover the cost. I was astounded when the Hospice nurse supported his advocacy. Two days later my father drove me to Denver where the neurosurgeon and a radiation oncologist began using their magical radiological weapon inside my brain.

My father and I kept this intervention a secret from Bob and Dr. Gonzalez. My silence was shame-based. My courage had faltered, causing me to succumb to expensive treatment in a desperate ploy to delay death. My father was silent because he worried that he had interfered in our marriage by subverting the Gonzalez treatment in which Bob believed.

I thought I would go to my grave with this secret. But others should know that in this instance, I opted for an expensive and risky surgical procedure in search of survival. The two lesions visible on the first brain scan had grown in size during the short interval between their initial discovery and the Cyber Knife treatment. The radiation oncologist was forthright that his treatments would not be curative. If nothing else, however, they gave my body time to push back against those perpetually re-generating, forever roaming, increasingly crafty and lethal cancerous cells.

-17-

Invest in Life or Prepare for Death?

The proximity of death raised risks in planning for a large celebration. The invitations had been printed, but not mailed. Bob and I were stumped. Should we sign a catering contract for Frank's celebration or prepay my cremation? Everything seemed a choice between investing in life versus preparing for death. In the end we decided to proceed with my father's eightieth birthday party; he certainly deserved to be honored for his longevity, steady devotion, and hard work. Also, since our wedding anniversary was three days before his birthday, it seemed appropriate to also celebrate what we believed would be our last.

There was another choice between life and death, whether to take a ten-day trip to Yellowstone Park. I had told Bob that if I was still alive in the summer of 1991, I wanted to take Eben to see Yellowstone. We had been anticipating the "bucket list" trip for the entire tenuous twelve months. If the neurosurgeon was right, we had nothing to lose. If Dr. Gonzalez was right, we could hedge our bets by hauling Gonzalez supplies into the park. We decided Yellowstone was critically needed targeted therapy for all of us, including Marley.

Sadly, only two days after we sent out the invitations, Bob's mother, Ruth, then seventy-seven, was diagnosed with breast cancer. She had a lumpectomy, followed by a lymphadenectomy. Nineteen of her lymph nodes contained cancerous cells—a number consistent with a pathologist's death sentence.

Ruth was baptized in the Church of Chemotherapy two weeks before my father's birthday party. While we assembled a cooler full of organic vegetables and fruits for Yellowstone Park, Bob's brother Joel and his wife Ellie flew from Los Angeles to Florida to be with Ruth. This was an incongruent reward for Ellie, who had just passed her own five-year checkup for breast cancer with flying colors. Breast cancer was a dark cloud following our family, even bigger than the one drawn by Addie.

We were loading our car with Yellowstone supplies when Charles F. Wilkinson, a close friend and former NARF colleague, stopped by to give us several of his well-known bear hugs, as well as something to read on our trip.

By nightfall we were in the midst of a thunderous light show as we crossed into Wyoming. As Bob drove, I read aloud by flashlight from the galleys of Charles' latest book, *The Eagle Bird.*[*] He had given us proofs for his chapter on Yellowstone—the descriptions of the geological transformations and the surrounding ecosystem were a perfect prelude to our visit. A few days later, the three of us were in a small boat in the middle of Yellowstone Lake, which, thanks to Charles, we were all acutely aware was the crater of a giant volcano. After catching two lake trout for dinner, we raced our rented motor boat wildly across the choppy water.

Sighting a bear among the lodge pole pines on the distant shore, I shouted to Bob and Eben, "Now this is living!"

After four glorious days in Yellowstone, we drove to the site of the annual Crow Fair at the invitation of Bob's client, Clara White Hip Nomee, the first woman elected to the lead the Crow Tribe. We sat outside Clara's tepee as she and her sisters-in-law graciously entertained us, along with the Montana Congressional delegation and other visiting dignitaries, using two camp stoves, Native wisdom, and their wits. With Marley at his side, Eben sat in the bleachers in a wide-eyed trance watching the Native American dancers circle the arena.

The day after we returned to Boulder, Aunt Alice arrived from Salt Lake, accompanied by her husband, Art, and her two married daughters—my cousins, Lynn and Nancy. Aunt Alice also brought two surprise guests, her ten-year-old granddaughter, Trina, (Nancy's daughter) and Trina's best friend, Callie.

At our dinner table, Aunt Alice's group was joined by Ruth, fresh from surgery and chemotherapy, as well as Joel and Ellie. It was a very informal carry-in dinner, the only kind of dinner party a Gonzalez patient can throw.

Before Alice and her family left to spend the night at a hotel, I asked her granddaughter Trina and her friend, Callie, both last-minute additions to the guest list, to write descriptions of each other for the name-tags I had been trying to create for every guest attending the party. Too exhausted to pay attention to what they had written, I added their notes to my secretary's file; she was coming in the morning to print the name tags.

---

[*] Wilkinson, Charles F. *The Eagle Bird: Mapping a New West.* New York: Pantheon Books, 1992.

The creation of the nametags was the only tug-of-war Olivia, Bob, and I had about the celebration. It was hard to conjure up clever, accurate, and enlightening introductions within a historical family context for each of the one hundred invited guests. Olivia and Bob were ready to abandon the effort, but, as the official hostess for the gathering, I felt anxious about the social mix. It was an extremely eclectic assortment of Jews, Mormons, Episcopalians, lapsed Catholics, atheists, agnostics, lawyers, scientists, Presbyterians, and psychotherapists. I thought the descriptive tags would help with integration.

Pictures of all the family and social groups in attendance were taken for framed post-party favors. For me it was a memorable catered celebration, tender and free of rain, as well as cooking, and clean-up. Most importantly, the gathering kept the dark cloud of death at an appropriate distance.

Eben with his paternal grandmother, Ruth Pelcyger, at his maternal grandfather's eightieth birthday celebration, Boulder, August 1991

Our closest friends were four other NARF couples, at least one of whom worked at NARF during the early years of institution building. After we no longer worked at NARF, we began having dinner together as a group, calling ourselves the "Original Non-Indian NARFers" or the ONINers. We were close in age, economic status, and political ideology. All of the ONINers and their children joined the celebration, which many knew was also a substitute for my anticipated funeral.

The ONINers, August 1991. Left to right: Richard B. Collins, Judy Reid, Bruce R. Greene, Susan Rosseter Hart, Bob Pelcyger, Joan Carol Lieberman, David H. Getches, Ann Marks Getches, Ann Amundson, and Charles F. Wilkinson

-18-

Strands of Synchronicity

The summer came to an end, but my life did not. Eben struggled to make a transition to his new school. At the end of his second day at Mapleton Elementary, he made two important declarations.

"I found a new friend, Mom. His name is Jesse Katz and he is really fun to be with," he said. "But, Mom, it's perfectly clear, I am the only kid in my class who can't read!"

Like all children, Eben had his favorite books, ones he repeatedly requested be read to him. During those repetitions, he had memorized the texts. The adults in Eben's life concluded that he had learned to read, blind to the fact that he was extremely dyslexic. I knew very little about dyslexia, but tackling the problem was a welcome distraction from my own disabilities.

In the middle of October, Bob and I flew east, combining his twenty-fifth Yale Law School class reunion and my regularly scheduled appointment with Dr. Gonzalez. It was the same week as the Senate confirmation hearings for Clarence Thomas to the Supreme Court, hearings starring Professor Anita Hill. Because I believed Professor Hill, my mind was filled with "Doubting Thomas" metaphors, making it even harder to not experience Dr. Gonzalez as the Wizard of Oz. After making a few adjustments in my program, Dr. Gonzalez sent us back to Colorado. Bob remained publicly hopeful, but I often caught him checking to see if I was still alive.

By November, I had developed what is called a "gait disturbance." My brain stopped sending signals to my left leg. I had a constant headache and was deeply depressed by the grind of treatment protocols—twelve enemas a day, two hundred fifty enzyme capsules, and other supplements to take around the clock, not to mention enough raw vegetable juice to make one gag. Then I had a dream about birds in Bozeman.

The Bird Dream

I had arrived for a visit in Bozeman, Montana. I left the hotel and went walking in search of things both familiar and new. I was accompanied on my walk by a bird that was unable to fly. We made slow, waddling progress. Above us, on a higher path, people were whizzing by—running, on skateboards, roller blades, and bikes. Above our lower path, in bare trees, sat menacing birds who resembled vultures. I felt very protective of the bird who could no longer fly. I came to a small apartment complex where I had lived for a year when I was fourteen and a freshmen in high school. I was surprised to see the buildings still standing and then noticed that there was a sign on one of the buildings which read "The Lie Low Club." A number of children's bicycles were parked outside, and, as I peeked into the windows, I saw what seemed to be some kind of after-school care center filled with children listening to music on headsets. They all seemed to be very happy and moving to the same rhythm. I felt a strong yearning to be among them. I walked on in the direction I thought the new high school had been built. Soon I came to the campus, and in the center of it was a three or four story red brick building. On the front of the building, students had painted a slogan. Suddenly, Eben was standing beside me and the bird was gone. It seemed very important for us to know what the slogan said, yet as hard as I tried, I could not decipher the words, nor could Eben.

I awoke in a state of preoccupation, my mind still filled with bird images. As I was getting dressed, the telephone rang. It was Judy Reid, the wife of Richard B. Collins. Judy is an ONINer well-known for her frankness.

"I'm calling," Judy said, "because I think the book that was given to me at Charles Wilkinson's fiftieth birthday party was meant for you! Ann must have mixed up the party favors when she put them by our place cards. It's called *Refuge,*" she said excitedly. "It's about birds and Utah and breast cancer. And the women in this family have the most wonderful conversations!"

At the ONINer's party for Charles Wilkinson, his wife, Ann Amundson, had given me a book about California history—Judy grew up in Berkeley. We agreed to trade party favors after she returned from a short trip.

Later that morning Roland and I discussed my dream—the flightless bird, my increasing difficulty walking, the easy symbolism of the vultures and death, my struggle to just give in to the demands of my treatment and "lie low." We also discussed my anxiety about abandoning Eben and his having to grow up without a mother—the difficulty both of us were having "reading the writing on the wall."

Walking home from Roland's office, I stopped at the Boulder Bookstore on the Pearl Street Mall to use the restroom. Preoccupied by birds and cancer, and too impatient to wait for Judy to return from her trip, I bought a copy of *Refuge—An Unnatural History of Family and Place* by Terry Tempest Williams.[29] I finished reading *Refuge* twenty-four hours later. Terry's poetic narrative opened my heart. The connections among the women filled me with tremendous yearning for my Nanie, for Aunt Mary, and for my cousin, Julie Ann. Terry's descriptions of her family and the birds of the Bear River Refuge were so beautiful and tender I read most of the text through tears. The deaths of Terry's mother and grandmother from breast cancer were strangely healing for me.

Terry believes that the plague of breast cancer, which has turned both sides of her family into a clan of one-breasted women, was a result of having lived downwind from the atomic testing area in Nevada. As I read, I found myself I mourning for my lost past and the less than *optimal distance* between my mother and me. Yet, the grief was mixed with tears of recognition. Perhaps sheer geography was responsible for my fate. If so, I could feel hopeful that my daughter might escape the legacy of breast cancer having spent only two days of her life in Utah.

When I finished reading *Refuge*, I noted Charles Wilkinson was listed in Terry's acknowledgments. I called Ann to ask her about Terry. She shyly admitted that she had bought *Refuge* for me, but had switched it just before my arrival at the party, worried that it might be too painful. The next day Ann kindly brought me other books Terry had authored.

On Thanksgiving weekend, as I was reading Terry's book, *Snow,* to Eben, I remembered she had referred to her niece "Callie" in *Refuge*. Surely, it was not possible that Terry's niece Callie was the best friend of Aunt Alice's granddaughter, Trina? The Callie who had arrived with Aunt Alice for the August celebration of my father's eightieth birthday? Trina and Callie would have been Gentiles to each other.

In the file for the party I found the final draft of the name tags for each guest. There was Callie as Trina had described her at my request the night before the party:

> *"Callie Tempest, best friend of Trina, traveled all the way from Utah. Trina reports that Callie is smart, funny, and pretty, and that she has good taste in friends."*

I retrieved the proofs of the group portraits from the party. In the photo of Aunt Alice's family, there was Callie Tempest standing to the right of my father with her hand on his shoulder. Strands of synchronicity filled me with a sense of inexplicable connectedness.

Frank V. Lieberman's Eightieth Birthday Party, Boulder, Colorado, August 1991 First row: Arthur W. Marshall, Alice Lieberman Marshall, Trina DeHaan, Frank V. Lieberman, and Callie Tempest with her hand on Frank's shoulder. Second row: Lynn Marshall Eaton, Craig Eaton, William A. (Bill) Marshall, Nancy Marshall DeHaan, and Vicky Taylor, friend of Bill Marshall.

-19-

Death, Memories, and Betrayal

Nuhiela Audeh died on Christmas Day 1991. Her *Thuja* triggered remission came to a sudden end and a new chemotherapy drug put her into kidney failure. The loss of her loving spirit was devastating.

In the middle of February, six weeks after Nuhiela's death, Aunt Alice and Uncle Art Marshall arrived from Salt Lake City to bid me goodbye. The proximity of death had led me to wonder what Eben would ever know about my own life given how little I knew about my father's. In an act of sublimation, taking advantage of the awkward silent spaces during the week-long visit, I asked my father and his sister to tell me what they remembered about their own.

For five afternoons, we sat together with tea and chocolates, while the two siblings recalled the memorable moments of their childhoods. No one has total recall of their life, so my father and his sister often disagreed about details—like the color of the coat their mother was wearing when she fell in a mud puddle on her way to work at Max Davidson's Columbia Cigar Company in Ogden, Utah in 1920. When my secretary transcribed the twelve hours of taped interviews, their narrative totaled one hundred fifty-three single-spaced pages, an invaluable record of Lieberman family history.[30]

During this period, Roland had been working with Jane Help, Margaret Clifford, and myself as a trio of dying mothers. We met with Roland to say what we couldn't say to those in our lives who were well. As in any group process, we also discovered hidden aspects of ourselves.

Sadly, it turned out that the biggest secret was being kept by Margaret's husband, Glenn Clifford. He had been having an affair with one of his patients. Learning that Glenn had been unfaithful to her for months, Margaret experienced appropriate temporary insanity.

Glenn Clifford's behavior highlighted the loyalty of Tom and Bob, but both Jane and I were now more acutely aware that we were at high risk for spousal abandonment. It was as if our clothes were suddenly sheer, revealing disfiguring scars which could not be mitigated.[31]

-20-

A Snowy Hurricane

On March 1, 1992, one year after the dream-like-appearance of the Being of Light, a heavy, wet snow fell on Boulder. The snow started at midnight and by dawn over two feet had accumulated. The weight of the snow damaged an astonishing number of trees. Boulder looked as if it had been struck by a snowy hurricane. Many streets were impassable due to felled trees and broken limbs. The cleanup effort lasted throughout March; piles of broken tree limbs on every block overwhelmed arborists, tree services, and trash haulers.

The piles of sticks and broken branches from the March 1, 1992 storm pushed me into sensory overload. Suddenly I was awash in nightmares and fragmented flashbacks. At first, Bob and I, as well as the Hospice nurse, and Dr. David Luce[32] thought it might be some kind of drug interaction. On March 9, 1992, I had a disturbing dream about my father.

Dreaming of Petrified Wood

> *Frank was very busy cleaning up my yard—filling a large dumpster to the brim with many things, but mostly with sticks and old pieces of petrified wood. Others, myself included, stood around watching, but we were not allowed to help with this task in any meaningful way. Frank was being very controlling about what went into the dumpster and how the sticks got stomped down. Then came an awareness that at some point my father was going to pour hot oil on the sticks and burn up the contents of the dumpster. I felt worried that this would cause trouble and wondered whether such burning was legal. But clearly Frank intended to burn it up regardless of the rules.*

The aftermath and remnants of the storm were what finally brought me face-to-face with the darkest aspects of my mother's paranoid schizophrenia. After the dream about my father and the stick-filled dumpster, I repeatedly dreamt of two dogs, and once about a children's book entitled *The Little Lost Dog.* These dreams led me to ask my father what had happened to Army and Nurse, our two springer spaniels in Delta. There was one photo of Army, several of both dogs in old home movies, followed by a sudden disappearance from the family album. My father's face lost color and he began sobbing.

"Your Mother made me shoot them!" he said. "She said they were taking after you. I hated to do it! Army and Nurse were good dogs and they loved you, but your mother was the boss."

The author with Army, Delta, December 1944

A man shooting his own dogs did not match the character of the man who built the feline dream house or the devoted hiking companion of Marley. There had to be more to the story. Suddenly I felt desperate to know everything before I died knowing nothing.

After walking Eben to Mapleton School on a Monday in the middle of March, I continued to my father's condominium. I had questions and didn't want interruptions. Leaving Marley by the side entrance, I buzzed myself in and took the elevator up to the second floor and knocked softly on his door. When my father opened the door he was still dressed in his nightclothes—striped cotton pajama bottoms, an old sleeveless undershirt, and leather slippers. Clearly surprised to see me, I explained I had come unannounced because I needed to ask him a few questions.

"I have a lot of questions about Mom—about when you realized she was mentally ill and other things I need to understand," I tried to explain.

My father moved into the kitchen and began to make himself a cup of tea. He started talking while waiting for the water to boil as I stood in the doorway of his galley-sized kitchen.

"You have always had a better understanding than me about Margaret's mental illness. When her Tucson doctor spoke to us after you committed her for the first time, I realized she had begun manifesting symptoms soon after your birth. Back then Bea Wallway told me I had to put your mother into an insane asylum, but I couldn't bring myself to do that to Margaret."

His tea kettle started whistling, but he continued, "It was hard to tell what was wrong because Margaret had a complete hysterectomy at the time of your birth. I don't know whether the doctor could have left her ovaries, but he didn't. The doctor thought their removal might prevent future asthma attacks because your mother's first asthma attack had been triggered by the hormones of pregnancy. Since he had to remove her uterus because of complications from the caesarean delivery, there would have been no further pregnancies, I don't think his judgment to also remove her ovaries was that good."

My father's voice softened when he said, "Margaret had wanted a boy and was quite adamant about her preference." Then he openly wept for a few minutes.

"We only learned she was going to have twins a few weeks before she went into labor, so we hadn't even readied the nursery for another baby," he said. "Margaret could not stop crying after learning your twin brother was stillborn. I guess the intensity of her hysteria caused me, and Nanie and others like her sisters, to avoid talking about your twin brother in front of her. As a result we never told you."

"When she was discharged from the hospital, Margaret was still so distraught that she wasn't capable of caring for you," he said. "So Nanie cared for both of you. On the advice of Nanie, before your mother brought you home to Delta, I hired June Davis, a young Mormon woman, who was training as a nurse, to help."

This startling narrative made me light-headed so I cautiously moved from the kitchen doorway to a chair by his dining table. My father, carrying his mug of tea, sat down across from me in his usual place.

"Look, Joan," he said "Margaret never wanted anyone to know that she couldn't have any more children. She was quite rigid and phobic about the whole topic. I am not certain whether this was part of her delusional thinking or whether it was her religious upbringing since many Mormon women believe their primary purpose is to bring as many impatiently waiting souls to earth as possible. Whatever it was, she was insistent that her condition be kept a secret."

"It was not long after Bea Wallway became pregnant with twins that your mother cut off contact with her. I wondered then if she wasn't feeling jealous. Later Bea told me she was certain your mother was envious.

"Peggy Stevens was Margaret's best friend growing up in Salt Lake. Peggy and her husband, Barr Quist, brought their son to Delta for a visit, but Margaret behaved so strangely she never heard from Peggy again, even though they had been best friends for many years," my father's voice was steady, but I could see tears behind the thick lens of his glasses.

"Bea was very concerned about your welfare," he said. "On several occasions both Bea and Marvin urged me to put Margaret in an insane asylum. Bea in particular felt Margaret wanted to harm you and that you should not be left alone with her. I was reluctant to do that and I found it hard to believe what Bea was telling me. The only option was the Utah State Insane Asylum in Provo, which was usually referred to as the Terrible Territorial Insane Asylum; it was well known to be a hell hole."

Throughout this part of his confession, my father seemed incapable of looking at me. Feeling dizzy, I asked him for a glass of water. My father brought me water in a wine glass, one of the glasses I had bought for him imagining he would have a real social life in his new condominium. The water tasted dusty.

"Is this the first time you have used any of the wine glasses? I asked.

"Yes, and I'm sorry I don't have any wine to offer you. I know that what I am telling you must be coming as a terrible shock!" While heavily disguised, this apology was extremely rare for Frank V. Lieberman.

After stepping into his bathroom to urinate, my father started talking again. "Remember when you decided to have a tubal ligation in 1972? The reason I asked you not to mention it to your mother was because I didn't want your decision to start another siege of reproductive paranoid anxiety in her."

"So, why didn't you just tell me the truth then?" I asked angrily.

"Well, I probably should have, but I have always tried to avoid conflict and it seemed better to let sleeping dogs lie."

He stood up and moved to his couch, perhaps to increase the distance between us.

"Look," he said, "your mother's illness has always controlled our communication. We both learned to do everything possible to avoid stirring up her demons. One thing I thought you should have been told before you were married was that you had only one ovary, but I let your mother decide these things, which I know was probably not the best thing for me to have done."

He stood again, moving even further away from me to his north windows where he seemed to be glazing at the cottonwood trees on Pearl Street. This was a hellish clarification conference, a technique I often advocated for clients in conflict or who had lost their way. Having opened this can of worms, I was feeling more furious by the minute.

"I remember being told I had only one ovary when I was pregnant with Olivia. I asked both you and Mom about it when you came to Berkeley right after her birth, but both of you literally acted as if you hadn't heard me! Why was that?" I asked.

"After your appendix burst, just before we moved to Bozeman, Dr. Budge discovered a tumor on your right ovary," he said. "When he told us, Margaret was extremely upset because Dr. Budge hadn't stopped mid-surgery to ask our permission. Instead, he had already removed the ovary. He recommended radiation to the area, but Margaret absolutely refused to allow the radiation, even though she had previously allowed Dr. Budge to radiate your thymus because you had grown six inches in all directions in four months.

"Anyway, after your surgery, your mother again insisted that no one be told anything except that your appendix had burst. Now I know that sounds crazy, but I didn't know what to do," he said. "Also your mother was already in a bad emotional state from having to sell the house and move to Bozeman. I didn't want to make the situation worse!"

My father confessor stopped talking to wipe his eyes and blow his nose.

My right hand was tremulous as I tried to make notes on what my father was saying. My bowels were also cramping, but I suppressed the disturbances so as not to break the continuity of my father's confessions.

"After Margaret fell and broke her leg in Tucson, the orthopedic surgeon told me she had severe osteoporosis—understandable since she was without hormones for most of her adult life. He had a heck of a time trying to repair the breaks and predicted she would have many more in the years ahead."

Back on the couch, my father continued, "The surgeon recommended your mother start on replacement hormone therapy, but he quickly became caught in her paranoia. Margaret rejected all his recommendations because he had put her back on Thorazine during her hospitalizations. I think the only reason Margaret agreed to travel to Europe after she was discharged was to get away from him."

My father and I spoke that day for almost six hours, while Marley waited outside with extraordinary canine patience. At age forty-nine, I was being told that I had been born a twin, but that my twin brother had not survived, and that my mother had a complete hysterectomy at age twenty-three and went the remainder of her life without hormone therapy.

I already knew I had only one ovary, but I previously had no idea why it had been removed. I had only the vaguest recollection of the thymus radiation, having remembered it incorrectly as part of Utah's Civil Defense program, along with the "atomic tattoo" of my blood type. Finally, I had completely missed all the obvious clues of my mother's osteoporosis.

My father's face showed utter exhaustion from speaking secrets. Since I had promised to meet Eben after school, I suggested we stop for the day.

My legs felt rubbery as I walked slowly up the hill on Eighth Street toward Mapleton Elementary. After a wave of nausea, I sat down on someone's rock wall in search of more stability. I didn't know if I was dizzy from cancer or dizzy from factual disorientation. Reaching Mapleton late, I found Eben in the Principal's office, with a look of concern on his face.

"Mom, I was scared you had died!" he exclaimed, making my stomach turn over again.

Walking toward Marine Street, with my son at my side jabbering non-stop about his day, I was deaf to his narrative. Inside our home, I helped Eben feed Marley before turning on the television to distract him with his favorite program, *Inspector Gadget.*

Marley finished eating quickly—she was tired from being on guard all day while waiting. She was ready for a nap. While Eben and Marley snuggled on the platform rocker watching *Inspector Gadget*, I poured myself a stiff drink, long-stricken from my extensive list of approved remedies.

I was astounded by how many secrets had been kept from me. I was in truth-shock, but I knew it was now necessary for me to re-frame my life.

Eben with Marley in the platform rocker watching 'Inspector Gadget' on March 15, 1992.

-21-

The First Witness

There were more dreams over the next several nights. With Roland's help, while under hypnosis, I was able to face some sensory memories from the day my mother attacked me before I managed to crawl out of the Chrysler at Topaz.[33] Most of the sensations produced such overwhelming feelings of fear, I couldn't hold them for long. The most persistent were the taste of fecal matter and the scent of vanilla; the only visual images were of sticks and rickrack. I desperately needed another perspective. I asked my father if he knew where the Wallway family was living. He did; they were residents of Casper, Wyoming, about three hundred miles north of Boulder.

When I telephoned Bea Wallway, she began to cry as soon as she realized why I was calling.[34] Pregnant with twins at the time of the Topaz incident, Bea had gone into premature labor three days after a Topaz Camp guard and the Millard County Sheriff brought me to her house. Bea said her twins were a boy and a girl. Her son lived only an hour; her premature daughter died twelve hours later. The sight of me had tortured her for years.[35]

After telling me more about what had happened, Bea said it would be easier for her to write me a letter. When it arrived, my last defenses were torn down. Bea even remembered she had used vanilla as perfume in those days of World War II rationing and that the two aprons she wore during that era were trimmed in white rickrack—symbols of safety I had unconsciously sought and had worn all the way to Africa.

March 24, 1992

My dearest Carol,

I apologize, but it will be hard for me to address you as Joan as I have always held you in my heart as Carol. Thank you for calling me and telling me about your life. I cannot believe both you and Mary Kay are battling breast cancer. Recently, in the tradition of the aged, I have been thinking more about the past. As I thought about the timing of the onset of both your breast cancer and Mary Kay's, I can't help but wonder whether there wasn't an environmental threat in Delta that was the seed of your tumors. Of course, Marvin's work with uranium and Frank's with DDT are logical suspects.

There is nothing but admiration in my heart for your attempts to understand the onset of Margaret's mental illness. Even though I was trained as a nurse, I had no way to make sense of what was happening inside her. You asked two questions that I will try my best to answer. First, here is what I remember about the day you were delivered to my door by a Topaz guard and the Millard County sheriff.

You were in a state of frozen terror, so obviously traumatized that the only thing I could think to do for you was to try to clean and dress your obvious wounds and provide you with a safe place to rest. Fortunately, when you first arrived Mary Kay was napping, so she didn't see you until after I had cleaned you up. Your condition when they brought you into our home would have been very frightening to her. Mary Kay didn't wake up until after I had cleaned and dressed your wounds and gotten you into a pair of her pajamas.

After all these years I still remember thinking it was good that Mary Kay was so much larger in size than you because I hoped it was less painful when I put the pajamas on, than if they had fit as tightly as they did on Mary Kay. You were absolutely mute the whole time I was bathing your body and cleaning and dressing your wounds, even though I know you must have been in considerable pain. I kept waiting for you to open your mouth and begin screaming, but you never did.

The sleeper you were wearing was beyond belief. I made a fire in the kitchen stove and burned it up. Only later did it occur to me that I should have saved it as evidence of what had happened to you.

When I gave you a cup of warm milk, silent tears started to roll down your cheeks. Those tears were the first sign that you were still capable of feeling and having emotions.

I felt very alone in my efforts to care for you. Both Marvin and Frank were out of town and unreachable. And your mother seemed to have disappeared into thin air. After you drank the milk, I made a bed for you on an army cot in the kitchen near the stove. You fell asleep almost immediately, even though Mary Kay had awakened and was trying to engage you in play.

I don't know why you weren't taken to Dr. Bird, except that it was a small town and you were such a mess that it was difficult to determine exactly what had happened. I do know that the sheriff had tried to find both your father and your mother. A neighbor across the street, Mrs. Ashby, suggested that I was a close friend of the family and the sheriff said she had also told him I was a nurse.

You didn't speak or cry the entire time you were with Mary Kay and me. Even when Margaret finally showed up, all you did was pick up her handbag and silently offer it to her. I am ashamed to admit that I have little idea what happened to you after she took you home. I do remember feeling relieved when Marvin returned and told me that Frank was also back in town. After I went into labor, I had to put my concerns for you aside.

As I confirmed on the telephone, in that era I did use an extract of vanilla in lieu of perfume. I still have a painful memory of your torn and soiled blue sleeper, as well as the aprons I wore while caring for you, both of which were trimmed in rickrack.

Second, you asked about Marvin's and my relationship with your parents after that devastating day. Our friendship did not end completely, but it was permanently damaged by your mother's behavior, even before the Topaz event. She had been increasingly withdrawn for what I recall was about two months. I was mostly bedridden at the time because of my pregnancy with the twins. Margaret suddenly stopped calling and coming to the house. Before then we had been spending a lot of time together, happily talking and cooking while watching you and Mary Kay play.

When I became pregnant with the twins and Dr. Bird advised bed rest, Margaret began walking over to our place with you in the stroller. She would spend time with both you and Mary Kay so I could rest. Then she stopped coming and she also didn't answer the phone when I called the house during the day. Marvin asked Frank what was wrong. Frank told Marvin that Margaret was afraid to sit on their couch because it had snakes in it and that only you could sit on it. Frank blamed her behavior on her hormones and told Marvin to tell me not to take it personally.

After I lost the twins, Frank came to the house. He offered his and Margaret's help in caring for Mary Kay. Despite my grief and shock, my instincts told me 'No' and I said I needed Mary Kay to stay with me. Only Frank came to the brief grave side services at the Delta Cemetery.

About a month later, Frank walked over to our house with you in your stroller and Mary Kay and you began to play as you always had. After that, arrangements were made by Frank and me to get the two of you together. I do remember celebrating your third birthday at their house—I was pregnant again which seemed to infuriate Margaret. Marvin and I left Delta as soon as the War ended.

After your parents moved to Logan and built the new house, there were two times that I recall seeing your family again. Once when we were returning from a vacation in California—I think you and Mary Kay were about ten years old at the time. The other time was when the three of you stopped to say hello to us in Casper on your way to Yellowstone. Both you and Mary Kay were then awkward teenagers.

Frank always wrote a lovely letter to us at Christmas time, and, for many years after we left Delta, I tried to remember to send you a birthday card. I wrote to your father several times over the years asking how you and Margaret were doing. His responses were always genial, but thin. I felt he was trying to keep me at a distance because I had repeatedly urged him to have Margaret committed.

He let me know when Margaret died, of course. Marvin and I always liked your father very much and we wish him well. Carol, if you are ever near Casper, I would love to see you again. Please know that you have a special place in my heart and that I will be praying for your continued survival.

With affection, Bea"[36]

The Wallway Family visiting the Lieberman Family, Logan, Utah, July 1951. On left: Frank, holding Lonna, and Marvin Wallway. Front to back on right: Patty, Donna, the author, Mary Kay, and Bea Wallway

-22-

The Second Witness

There was a second witness. He was home alone on February 20, 1945 because his face was covered with chicken pox scabs. Drawn to the front window by the sound of the wind cracking a tree limb, he pressed his tongue to the cold glass and squinted to see across the street through a raging sand storm. One of the giant elms standing sentry in front of our house had split open—one half falling onto the front steps, the other half boughs had been dismembered. Broken branches from other elms flew through the air like confused arrows.

The storm was so fierce the witness began to imagine the trees being uprooted and falling across the road toward him. He tried to measure in his mind whether the top-most branches would reach the window where he stood. Luckily he was in Delta, Utah—a community where the streets were wide, designed by the Mormon prophet Brigham Young, whose followers believed he was using God's blueprint. The extra width allowed for turning a team of four oxen pulling a covered wagon. Due to prophetic planning, the witness was watching at a safe distance.

Forty-seven years later, Roger Ashby remembered my mother was wearing high heel pumps that day. He caught sight of her coming out the side door of our house, just as the wind whipped her skirt above her waist. I suspect he was infatuated with my mother in the way pubescent boys tingle in proximity to a beautiful, reportedly dangerous woman.

He was frozen by the sight of her struggling to keep her skirt down with one arm, while pulling me behind her down the side steps with the other. He watched her drag me by the collar of my sleeper toward the shed while our two springer spaniels jumped wildly around us. Quarantined by both his illness and his fear of spiritual contagion, Roger Ashby was too afraid to come to our aid. My mother and I survived. The two springer spaniels did not.

I found Roger when I was fifty and he was fifty-six. I wanted to follow up on the "Mrs. Ashby" that Bea Wallway had mentioned in her letter. The only childhood memory I had of anyone named Ashby had been formed by an early photograph, a snapshot of me with two other girls and two boys sitting on the cement steps of a house. Behind us a large window was divided by mullions. Written on the back in my father's hand was the notation, "Carol with Annie Miller and the Ashby Brothers, December 1946." The picture was taken almost two years after Roger Ashby stayed home alone from school with the chicken pox.

My search led me to telephone the Delta City Library. The librarian provided me with an instantaneous answer, not from library resources or the Internet, but from the collective memory circuits that often exist in small towns. She identified the correct Ashby family of many in the area during that era.

It was the ultimate cold call. I dialed his number on Sunday, March 29, 1992, in the afternoon. When he came to the phone I hesitated, my mouth suddenly as dry as if I were speaking to a large unknown, potentially hostile, audience.

"I am sorry to disturb, you, Mr. Ashby. My name is Joan Carol Lieberman and I believe my family lived across the street from your family in Delta. It was when I was little girl in the early 1940s. I was wondering, do you have any recollection of my family?"

His memories were sparse, but vivid: developmental fragments, strands of images, living frames that had been frozen and stored away in the peculiar preservatives of childhood memory. It was as if he had been waiting all those years for me to reach out and ask him to retrieve his secret specimens. After giving me permission to tape our conversation, he started with the rules of social engagement in the 1940s. Among them was the fact that my parents were strangers in Delta, which then had a population of about one thousand people, and my parents didn't go to the Ward house.

"We heard your mother came from a big Mormon family in Salt Lake, but everyone knew right away from his name and his looks that your father was something else. Your mother was so pretty and with all that red hair she drew a lot of attention. Then when you were born and she didn't bring you in for a blessing or nothing, nobody quite knew what to do about you. I had been told to keep my distance and did for the most part. But I admit your family was the first I knew that were Gentiles. I was supposed to see something kind of devilish in you and I guess it was my nature to want to see something like that. So I spent a lot of time watching your house."

"It was confusing to me," he continued, "because one or more of you always seemed to be disappearing. I got to know your Dad after your family got the dogs because we didn't have a dog and I was of the age when I wanted one real bad. Your Dad must have seen it in my eyes or something. He was always bringing those spaniels to the fence so I could pet them and say hello."

The author in front of her family home surrounded by elm trees, Delta, December 1944

Our conversation that day lasted over an hour. It was toward the end that I asked him if in all the watching he had done, had he ever seen anything really strange.

"I mean, anything that stood out from our normal strangeness?"

Then his memories of that day came spilling out like a wound in need of lancing. He started with being home alone, his tongue on the cold window, and the storm. He was alone because he had no fever so his mother had gone over to the Ward house. He was at the stage of the disease where they were waiting for his scabs to fall off, so he could go back to school. It had been a long siege; his twin brother and sister had come down with the chicken pox first, and finally he had been stricken.

"Those dogs were menacing your mother. I could tell that for sure. When she went after you with one of those broken branches, the dogs backed off a little distance—like they was expecting her to come after them. I had a hard time imaging that you could have done anything so bad as to deserve what she was doing to you with that broken branch. When she finally left you alone and got into the car and started to drive away, I could tell those two spaniels were trying to help you. They were licking you and staying right by you."

"Then your mother backed up the car real fast and got out. When she started toward you, those dogs went crazy! She had to pick up the big branch again to keep them off her. I remember she was holding you real strange, at arms' length by the back of your sleeper. Your arms and legs were dangling down like you were a kitten or something. When she got you close to the car on the street, I could see there was blood and stuff on your back side and all over the sleeper. That sleeper was a strange blue-like color—kind of like the color of a robin's egg. Every time I have seen that color for the last fifty years, a picture of her holding you like that has crept into my mind.

Those dogs were trying to leap over the fence pickets and I thought for sure they were going to make it. Your mother got something out of the trunk and wrapped you in it real tight. Then she put you in the back of the car where I couldn't see you and sped away."

My mouth seemed stuffed with cotton and my stomach was heaving, but I knew I needed to hear whatever he remembered.

"The next morning," he continued, "when your family car was still gone, and there was no sign of your dad's truck or sprayer rig, I went over and fed the dogs after my mother went shopping. The dogs had been barking half the night. I could see from the window that they were waiting and looking real hungry. I just opened the back screen door to your house and saw their food and bowls there. I fed the dogs two more days before I saw the car was back.

"I tried to tell my mother about what I'd seen, but she just yelled at me about keeping away from them Gentiles, telling me if I didn't my Pop's belt would be on my back. I think it was about three days before I saw your mother pull up in her car. Later she left and came back with you. I think it was the next day that your dad came back. Then the spaniels disappeared. Just like that!"

"There is something else I remember," he said. "Some weeks later, when spring had come on, I asked your Dad about where the dogs had gone. I still remember he started to cry. He was the first man I ever saw sobbing. His face always stayed stuck in my mind. Then, when all that stuff about Jews and the extermination camps came out, I thought that was maybe why he was crying. He never did give me an answer about the dogs though."

The author (on right, first row), with Annie Miller, Ashby siblings behind; Roger Ashby is on right behind author, Delta, Utah, December 1946.

I told him I had a small snapshot from that era that showed us together with a girl named Annie Miller.

"Do you have any idea who that might have been?" I asked.

"Oh, yeah, I know!" he said. "Annie was my brother's little girl friend. We talked her into crossing the street and inviting you over to play so that we could take your picture. The Bishop had talked at Testimony Meeting that week on the dangers of Gentiles. He said that the Devil's people didn't show up in photographs. My mother even agreed to take the picture.

"When the picture came back from Salt Lake and you were in it, my mother told me to get it out of the house. So I took the photo over and gave it to your Dad."

-23-

An Inexplicable Remission

What happens after death is an impenetrable secret kept from the living. All I know is that after a close encounter with death, it can be a challenge to resume living. As medical technology and new pharmaceuticals bring more of the terminally ill back from the brink, there will be many more questions about how to live successfully as a surprised survivor. While struggling to stay alive, never once did I worry about what would happen if I didn't die. My rational mind seemed ill-equipped for survival.

On Friday, April 17, 1992, I went to the hospital for an MRI of my liver and brain. As I maneuvered myself onto the scanning platform, I told the technician "I don't think you are going to find anything." We both assumed my comment was the small talk of a nervous patient.

Later that evening, Bob, Eben, and I sat down at our dining table to share a modest dinner. We lit candles and poured a glass of wine for Elijah because it was coincidentally both Good Friday and Passover. These were not normative rituals for our faithless household, but something about the day seemed sacred. While we were still at the table the phone rang. It was Dr. David Luce, the internist who had agreed to follow me locally after I began the Gonzalez protocol. Dr. Luce called to tell us my scans were completely clear. I had somehow won the lottery of life without even buying a ticket!

The source (or sources) of my temporary remission remain a mystery. Dr. Gonzalez immediately claimed credit, which both Dr. Luce and Bob readily ceded to him. My own feeling is that whatever happened in my body was a complex and mysterious process. I still hold my remission in a sacred place. I was expected to be happy, grateful, and full of zest, but I found myself instead in a crisis of faith. My scientific ethos had been blown to bits.

Three weeks after my radical—or some might say miraculous—remission, I kept my regularly scheduled appointment with Dr. Gonzalez. I traveled to New York City in a state of great ambivalence because he was still a dark figure in my psyche and I was uncertain how he would appear in this new light. Sitting in his waiting room, I had to force myself not to leave. When Dr. Gonzalez came to escort me into his office, he told me a physician colleague interested in his work was visiting from Holland.

"Do you mind if he observes your consultation and exam?" Dr. Gonzalez asked.

I nodded permission, following him into his office, where a tall white-haired gentleman stood waiting—one whose powerful presence manifested elegance, intelligence, and caring.

Dr. Gonzalez introduced him, "This is my colleague, Dr. Hans C. Moolenburgh, a Dutch specialist in non-toxic cancer treatments."[37]

Dr. Moolenburgh graciously thanked me for allowing him to observe and the three of us took our seats. I knew Dr. Luce had already sent my films to Dr. Gonzalez, so I wasn't surprised when he made his request.

He said, ""I want to ask your permission to send your tumor samples, films, and case records to the National Cancer Institute (NCI) at the National Institute of Health (NIH). Do I have your approval to do so?"

"I guess so, as long as there are no other requirements," I replied.

I knew Dr. Gonzalez had been seeking approval for a clinical trial of patients with pancreatic cancer, which is extremely lethal. Part of me was curious to see those results and part of me was worried about being swept up into an endorsement of his work. He gave the report and films of my most recent heart study a cursory glance, before he launched into his review of my blood and hair test results, a meaningless numerical narrative. When he finished, Dr. Gonzalez asked if I had any questions before we talked about the changes he was making in my program.

"Yes, I do," I said. "My greatest problem is that I have no framework for understanding my remission. It doesn't seem possible, particularly given the extent of the metastatic disease in my liver and brain."

Before Dr. Gonzalez could respond, Dr. Moolenburgh leaned forward, and, in perfect English, with a barely discernable Dutch accent, spoke in a most compelling manner.

"I have reviewed your records carefully, Miss Lieberman," he said. You must believe in your own miracle!"

Dr. Moolenburgh's words, the sound of his voice, his warm hand placed lightly on my forearm seemed to melt something that had been frozen inside me. It was as if I was Manx meeting Bob for the first time. My defenses and doubts dissolved.

As I exited the office to go into the exam room, I saw a book entitled *Meetings with Angels*[38] by the door. The book was incongruent with the plethora of file folders containing patient records and over-sized green file jackets filled with film studies that covered every surface in Dr. Gonzalez's habitat. During the physical exam, I found Dr. Moolenburgh's observations precise, more insightful than those of Dr. Gonzalez. Returning to the office, I glanced at the book again and saw that the author's name on the cover was Dr. H. C. Moolenburgh.

Looking at Dr. Moolenburgh, I asked, "Is this your book?"

"Yes, it is," he responded with a soft voice.

A short time later, I said goodbye to the two doctors, paid my bill at the front desk, and walked out onto Park Avenue. The sun had disappeared behind the tall edifices; the irritable chorus of honking cabs and cars was jolting. The image of Dr. Moolenburgh seemed surreal. Had there really been a book about angels on the table? I couldn't even seem to remember the title or the Dutch doctor's name. I buzzed myself back into the office to ask Dr. Gonzalez's assistant to write down the doctor's name with the correct spelling.

Back out on Park Avenue, I hailed a taxi to Penn Station where I took a train to Huntington Station on Long Island for an overnight visit with Olivia. The Long Island Railroad environment was so devoid of the kind of feeling evoked by Dr. Moolenburgh's presence that I kept my eyes closed.

As I waited for Olivia to pick me up, I tried to reclaim my mother persona so that I could be in closer contact with my daughter's scientifically-oriented atheistic self. It was a struggle to put aside the feeling of emotional vulnerability Dr. Moolenburgh had stirred in me. After I climbed into her car, Olivia asked me how the visit with Dr. Gonzalez had gone.

I surprised myself, saying, "Fine! He had an angel of a man with him!"

Because it sounded like something my mother would have said, Olivia asked no more questions. She had years of training in not following up on such statements.

Addie and Katherine Help with the author after her radical remission, September 1992

-24-

Denevere's Doula

No chemotherapy or radiation was proposed for Jane after the spinal surgery in the fall of 1991 because her surgical wounds were too extensive. Nor am I certain that Jane ever took another tablespoon of *Thuja* after her second surgery. Mostly, she enjoyed Dryers ice cream and an occasional piece of pizza delivered by Domino's. As if her rented hospital bed was still housed in its institution of origin, Jane kept the television on continuously.

Every morning except Sunday, Tom left home before daybreak to turn on the machines at his business. Later the children would eat their breakfast cereal sitting on the floor or on the couches near Jane. After the older children left for school, only Adam remained at home until Jane's mother, Jeanne, arrived mid-morning. Then seventy and a widow for fifteen years, Jeanne had moved from California to help Jane after her first surgery.

Jeanne helped Jane with her personal care, chatting nervously to obliterate the anxiety she felt whenever she glanced directly at the destruction of her daughter. Often after these hygienic efforts, Jane drifted into a morning nap. Adam would also take a much needed nap, having taken possession of his mother's previous bed, one of the crumbling chenille couches, only occasionally sleeping with Tom on the water bed. Jane's pain was now so great that there were only brief snuggles. Adam had to settle for proximity in lieu of the warmth of her body, sleeping only when she slept, as if not wanting to miss a moment of his mother's remaining existence.

Car pools were organized, as was a schedule for a meal brigade of mother-cooks from the Friends' School parent group. Haphazard stacks of aluminum casserole pans and empty Tupperware containers rose in a precarious tower amid the bikes, tools, and other garage clutter.

Jane slowly recovered enough strength to tolerate a ride in the Chevrolet van, which Tom modified to accommodate her wheel chair. Tom and Ronald removed the middle row of seats so Jane's wheelchair could be secured with jerry-rigged seatbelts. A Sunday drive in the mountains became a family narcotic, the familiar movement suffusing what lay ahead.

Jane had another year of unexpected survival. Half-way through the year, Dr. S attached a pump to Jane's port to administer a small but steady dose of FU-5, which, in the doctrine of the Church of Chemotherapy, was believed to prevent a terrible end. With the continuous infusion of FU-5, Jane became increasingly serious about getting around with her walker. Her arms had remained strong from the use of the overhead hoist attached to her bed and her will to move independently was indomitable.

On Mother's Day 1992, a few weeks after my own radical remission, I was downstairs doing laundry when Marley announced visitors. As I made my way toward the front door, I was stunned to see Jane standing upright, slowly moving up our front walk with her walker, smiling a cheek-breaker. Jeanne, also grinning and carrying a bouquet of flowers, was directly behind her.

Whenever I hear a negative cancer prognosis, I remember a smiling, upright Jane moving toward me in the May sunlight. There was no explanation for Jane's recovery in medical or neurological science. And like many modern day medical miracles, the surgeons and oncologists handling her case seemed to avoid speaking directly about it. What is it about medical training that turns doctors silent, suppresses their curiosity, ignores anecdotal evidence, and denies possibilities?

By the end of the summer Jane had recovered enough movement to graduate from the walker to a cane and was once again able to drive, perhaps to do some secret shopping. Jane called me from a pay phone outside Kmart, laughing and weeping with joy for her reclaimed independence. It was a rare moment of *optimal distance* for a woman who had miraculously broken the bonds of her terrible tumors. For the next five months Jane drove around Boulder fueled by morphine; she was a more competent and attentive driver than most in a city heavily populated by out-of-state college students, mindlessly driving over-sized SUVs.

Becoming a breeder of Pulik was one of hundreds of small business ideas Jane hatched during her lifetime. The superior intelligence of Marley and Duncan, coupled with the difficulty of finding a Puli, had convinced Jane there was an unmet market for Hungarian Pulik. She decided to acquire a female mate for Duncan. Her children named the female "Denevere." On the Fourth of July 1992, Denevere gave birth to six puppies, four females and two males—Jane's anthropomorphizing act of sublimation. The children gave each puppy a name that began with the letter "D" and each night their assigned puppy shared their bed.

In September, Jane loaded the puppies, along with Denevere and Duncan, into the Chevrolet van to pay a surprise visit to Adam's class at Friends' Preschool next door to our home. The moment the van parked on Marine Street, the children and Pulik puppies became a giggling, licking mass of energy and love on the front lawn. Denevere sat in the middle, growling an occasional warning whenever a child or puppy was being too rough, while Aunt Marley and Daddy Duncan kept guard on the sidewalk perimeter.

When they reached the age of adoption, I persuaded (really put extreme pressure on) the Getches and Amundson-Wilkinson families to adopt a puppy. David and Ann Getches took Dakota, while Charles Wilkinson and Ann Amundson gave Dusty a home. Jane's last summer on earth was defined by the pleasure she took in being Denevere's doula.

-25-

"Running a Banana Plantation for Jesus Christ"

The moment death passed me by, it took aim at my father. Seventeen years earlier, my father had undergone open heart surgery—a quadruple by-pass operation in Salt Lake City. The surgery had taken place two weeks after Bob and I were married. Before and after that surgery, my father had lived the life of a dietary monk[39] trying to keep at bay the heart disease that had taken the lives of his father and two older brothers.

For twenty-five years Frank had poached his boneless skinless chicken in water, made his homemade whole wheat bread without oil, and steamed his potatoes and broccoli. On rare occasions, he had a piece of dark chocolate. Otherwise, Frank V. Lieberman had led a fat-free life since 1969, hiking through the Rocky Mountains with Marley at least ten miles every week.

Within a few weeks of my remission, my father began experiencing angina. Dr. Turvey ordered an angiography. Despite his determined dietary sacrifices and exercise rituals, the arteries of my father's heart were once again almost totally occluded. At the age of eighty-one, my father was told for the second time that he needed emergency open-heart surgery. I rode in the ambulance with him from Boulder Community Hospital to University Hospital in Denver. We were met by an extremely gracious and thoughtful senior cardiologist, who broke the medical mold with his first question.

"What is your profession, Mr. Lieberman?" he asked.

At five in the morning, the University Hospital team began cutting open my father's chest for the second time. Frank had been in the operating room for twelve hours when the same senior cardiologist approached me in the surgical waiting room.

His speaking demeanor was tender, "I am deeply sorry to tell you the surgeons have been unable to get your father's heart to restart. They have been trying for several hours to wean him off the heart-lung machine, but they have been unsuccessful. Do you wish to see him before the life-support equipment is disconnected?"

Speaking through tears, I remembered aloud, "That's the same problem the surgeons had at St. Mark's Hospital in Salt Lake City seventeen years ago."[40]

The senior cardiologist returned to the operating room with this nugget of information. After six more hours of waiting in agony, I was taken to see my father in the recovery room. He was on a ventilator and a balloon pump had been placed in his groin to supplement his weak heart. His left arm had been sliced open because the vein where an IV had been inserted had blown.

My father remained in the ICU for six weeks. At the end of the second week the ventilator was finally removed. My father's first words, spoken in a low raspy voice, filled me with mirth.

"I'm sorry to have been gone for so long, but I have been away in South America running a banana plantation for Jesus Christ!" he said.

Frank V. Lieberman was utterly perplexed by my inability to stop laughing. During my daily visits over the next several weeks, my father, as if he were my mother, repeatedly besieged me to go after the large cockroaches and ants crawling up the wall of his ICU cubicle. Playing along, I pretended to vigorously dispatch them.

In June, when my father's thought processes and lung function were more normal, he was moved to the cardiac step-down unit. He was finally released to my care in July and a hospital bed was installed in our living room. Every day for six weeks, I nursed him—duties that included twice daily saline washing of the open wound in his groin, the former location of the balloon pump. The incision had to heal from the inside out. This intimate aspect of nursing my father was definitely not my idea of *optimal distance*.

During my service as his nurse, I noted something curious—my father had been circumcised! Because he would have been far too young to remember the circumstances, I did not ask him. However, I did wonder whether he was born with a "natural circ"—a genetic trait he may have passed on to his grandson.[41]

-26-

The Bear and the Trumpeter Swan

My stint as my father's nurse ended in September. Bob's travel was in a rare lull, and Eben was back in school, overcoming dyslexia, following a celebration of his eighth birthday. October seemed to offer a rare break in the dark clouds of responsibility that seemed to follow me everywhere. In search of solitude, I drove north to Yellowstone National Park, a place providing protection from the needs of others.

I felt almost giddy in a red rental car on the back roads of Wyoming. My plan was to enter the park through the East Entrance, so I drove five hundred miles until I reached Cody. I took a room at the Holiday Inn and awoke to the first snow of the season.

In the morning, the desk clerk warned me that the East Entrance road was closed due to the snowfall. Since the sun was shining, I drove toward the park entrance anyway, hoping it would open later in the day. The Park Service roadblock was still in place when the sun started to set, so I took a room at a nearby dude ranch. The only other guests in the deserted dining room were four radiologists who had come to hunt elk on horseback. I listened to their table talk of tax shelters as I ate alone at an appropriate distance.

On the second morning I drove over Sylvan Pass toward Yellowstone Lake. The radio weather report indicated the day was likely to be the best of the week, so when I saw the trailhead sign for the Thorofare Trail, I pulled off the road and parked. Bob, Eben and I had hiked a short portion of the Thorofare the previous summer and I decided to retrace our steps. After putting on extra socks, I added the dude ranch sack lunch to the minimal contents of my back pack, and walked into the deep forest.

The cedars were dripping melting snow; the smell was sharp and sweet, fresh and ancient. A million sun signals led me through the shadowy trees. When I reached Cub Creek, I bushwhacked down to the shore, just as we had fourteen months earlier. A nearly invisible path, perhaps ours or ursine, crossed back and forth over the mossy boulders and fallen trees in Cub Creek. As the forest opened to the beauty of Yellowstone Lake, my eyes filled with tears.

Reaching the shoreline, I was surprised to discover I was bleeding. Was it possible I had begun to menstruate again? This was a phenomenon younger women sometimes experienced after chemotherapy, although their eggs were sterile. Unprepared for such an event, I tried to ignore the evidence. Feeling ravenously hungry, I quickly ate my Wyoming Wonder Bread roast beef sandwich and apple. Then I lay down in the sun because, despite the cool temperature, my body felt wonderfully warm from the combined power of sun, snow, and water. No one else had signed the trailhead log; I was alone and at peace with the natural world.

After my warm rest, I decided to strip below the waist and submerge my lower body in the waters of Yellowstone Lake—both to cleanse myself and to self-baptize my mysteriously healed body, an unusually self-conscious act on my part. Removing my boots, socks, slacks, and underpants, I gingerly made my way into the cold clear water. When the water reached my knees, I turned halfway around to face the shoreline so that the sun would warm my back. Then, holding up my shirttails, I lowered myself into the water, watching it turn briefly pink. After several waves lapped at my body, only the soft colors of the stones on the lake bottom could be seen.

As I was moving out of the freezing water, feeling the sun on my flanks, I heard a disturbance to my left. Turning toward the sound, my eyes met those of a huge brown bear standing in the tall grass along the shoreline. About twenty feet behind her stood twin cubs. The ursine trio was less than one hundred feet away.

One glance produced the sensation of warm urine on my thighs. My legs turned to lead as I struggled toward the shoreline. I knew the rule was to slowly back away while trying to look as large as possible, but I would die of hypothermia if I retreated back into the waters of Yellowstone Lake. Momentarily paralyzed, I managed to grab my pack and boots, before I turned right and started to run east along the shoreline. Ineffectively trying to protect my bare feet by leaping from one sandy spot to the next, even as the flow of blood from my uterus coated the inside of my thighs. Moving required enormous effort. My body felt as if I had run for miles, but I had only gone a few hundred yards. As my lungs reached their capacity, I let go of my pack, then my boots—one by one. I could hear the bear gaining on me. The ironies of this kind of death were too much.

"Not now! Not here! Not this way!" I was surprised to hear myself screaming, as if there were someone to listen. "Please, it will be too hard on Bob, Olivia, and Eben. Please!"

The bear was so close my nostrils burned with her scent and filled my bronchial tubes—now in violent spasm. What I remember most clearly about those moments is how my psyche slowly moved into a state of acceptance, not unlike the sensation of morphine or another powerful narcotic moving through my body's tributaries. It would not be the worst kind of death, certainly better than a hospital or hospice death. My body would serve a natural purpose, becoming a meal flavored by irony.

In this enveloping state of slow-motion submission, still struggling to take each step, I became aware of a mysterious sound, not unlike a stiff straw broom moving over rough flagstones. My ragged breath seemed to fall into the same sweeping rhythm. Out of my right eye I saw a huge white bird descending from the sky, coming toward me like an arrow. I raised my arms reflexively in an effort to protect my head, but I stumbled, lost my balance, and fell. My left knee struck a rock at the same moment a powerful swoop of feathers struck the right side of my head. The bird's body hit the ground behind me and a mournful, plaintive sound seemed to pierce my body.

Turning halfway around, the mother bear's right paw had pinned the bird by one of its huge wings, while the other wing flapped wildly. The bear's jaw opened wide before her yellowed cuspids crushed the long neck of the bird. After an animal moment, her ursine jaws opened again, quickly closing again over the bird's body. Our eyes still locked, the bear turned, carrying the now limp bird back toward her cubs, still standing like twin sentinels.

Shock battled awe. Every cell in my body was in motion. My hands shook, my chest heaved with sobs. Struggling to stand, I staggered further east along the shoreline, coming to rest behind a pile of washed tree flotsam and a large boulder. There on my knees, I gave thanks to the universe. A long time passed before I was able to reclaim enough breath and equanimity to move.

As the sun moved lower in the sky, turning the mountains of the western horizon a deep purple, I cautiously made my way back toward my boots and pack. I had watched the mother bear and her cubs move west, away from my stone refuge in the direction of the outlet of the Lake. My feet were badly cut and bleeding; my knee beginning to swell. I could see my corduroy jeans in the distance, but they were further west than I dared venture. Retrieving my boots and pack, I circled wide, up to the edge of the forest. There I made my jacket into a tight short shirt to cover my lower body and created a pseudo-tampon from Kleenex in my pack.

Then the curious scientist in me, along with the child who kept her treasurers in a Purple Satin Hanky Box, took my still shaking body by each hand, and the three of us moved into the tall grass. One white wing lay on top of a fluff of bloody down on the wet grass spotted with snow. Gently picking up the wing, I placed it across my chest inside my shirt before making my way, eagle-eyed and eared, along the east side of Cub Creek to the Thorofare Trail, leaving my blood stained corduroys behind.

Inside the car, I cleaned my body as best I could with hand wipes before putting on clean underwear and slacks. I was still shivering so hard my teeth sounded like they were talking to each other. With the car heater on high, I drove slowly west along the Lake, turning south for a few miles, and then west again toward Old Faithful Inn. The Inn was booked solid with elderly tourists from Florida, their Tauck Tour buses lined up like a long train.

With bandages, hydrogen peroxide, and tampons from the gift shop, I used the lobby restroom to wash my body more thoroughly with brown paper towels and liquid soap from the dispenser before dressing my wounds. Too tired to drive out of the park in search of another room, I slept cold and crunched like an injured fetus in the backseat of the red rental car, dreaming of strange wild bear-like beasts, birds, and deep, dark lakes.

In the morning, I ate a restorative hot breakfast in the dining room of Old Faithful Inn. I sat alone, surrounded by the collective unconscious of Florida, their voices a chorus of New York accents repeating the same meaningless refrains about their preferences for items not on the Yellowstone menu.

After breakfast, I walked from the Inn to the Visitor's Center to ask the ranger on duty if he could identify a bird from one wing. He couldn't, or wouldn't, but he sent me to another ranger, in an unmarked building. The second ranger hadn't even held the wing before he announced it was from a trumpeter swan.

"I am guessing that it belonged to a female, probably one who had lost its Cob since the main flock left the Yellowstone and Madison River several weeks ago." He seemed confident and knowledgeable.

The ranger started to tell me I had to leave the wing with him, but he must have seen something in my eye because he quickly relented.

Still in shock, I felt desperate for a spot that was safe, warm, and private. The Old Faithful Inn was fully booked for the remainder of the week. Driving along the Firehole and the Madison Rivers, I felt as if I was floating away on the late autumn ribbons of water. Passing the gaggle of empty souvenir shops in West Yellowstone, I thought of Jordan, remembering how his sweet face looked as he stood next to the grizzly bear.

There was no lodging in West Yellowstone, so I turned south on Highway 20 to Ashton, Idaho. Wanting to hug the mountains, I made dust clouds on narrow unpaved back roads to Driggs and Victor. I felt as if I was driving through fields of gold, but they were just fields of cut hay lit by sun from the west, deepened in their golden color by the shadows of the Teton Range. I drove over Teton Pass in the dark, snow blowing hard at my windshield, and took a room at Teton Village.

Drugged by dreams, I slept deep into the morning. After my second hot shower, I ordered a box lunch from the hotel restaurant. I drove a short distance to Jenny Lake and hiked around its perimeter. In the last rays of sun, I sat on a boulder at the edge of the shore. Soon a chickadee approached asking politely for a few crumbs. I opened my palm, and the chickadee hopped on my hand to eat the sesame seeds I had picked off the crust of my sandwich. Perhaps this particular chickadee was only an experienced beggar taking advantage of one of the last lunch opportunities before heading south for the winter. Regardless, at that moment, I felt as if we were the same species—two lonely creatures savoring our last moments in the North together.

For many months after I returned from Yellowstone, I was completely preoccupied with the bear and the swan. Yet I was reluctant to speak about my experience, even to those closest to me. Habitually, normatively, I feared being mistaken for my mother. More importantly, I also feared a dilution in any discussion. Mine was a metaphysical experience beyond belief, demanding faith, instead of rationalized possibilities.

Since that day, usually at moments when I am feeling utterly alone, I find myself wondering if it was my mother who saved my life. My rational mind knows this only proves that my thirst for maternal reparation will never be quenched. I also wish to report that daily living produces a dust of forgetfulness, which slowly turns into uncertainty. Nevertheless, there have been a few moments when my defenses, masquerading as doubt and denial, have been stripped away. At those moments, I believe there are spiritual forces of great strength, intelligence, and love in our universe. This I know because I once experienced their palpable presence.

Bear cradling swan sculpted by Bob Epstein for the author in 1994.

-27-

On Being Death's Doula

Cancer took control of Jane Help's body in late October. As Jane began her final labor, her hard-earned mobility was washed away by tidal waves of agony. The FU-5 chemo pump was disconnected. In an effort to control her pain, a small incision was made to cut the nerves the surgeons had so carefully tried to preserve, as well as those Jane had miraculously regenerated.

Jane and Tom had communicated minimal information to their children about their mother's illness. Children sense everything, but the older ones, better at translation and observation, knew more. Tom finally garnered the courage to tell Addie and Katherine their mother was going to die as they sat on my lap in the laundry room. My thighs felt excoriated by their pain.

Sobbing through her grief and outrage, Katherine cried out, "It isn't fair! It just isn't fair!"

Jane never did write the Hospice-promoted letters to her children. So much of her mothering manifested in a glance or a gesture, it was neither possible, nor appropriate, for her to change modalities.

It had been difficult for me to persuade Jane to let Hospice in the door, but the nurses hired through the hospital service were either overwhelmed or ill-equipped, sometimes both. After the fourth nurse quit in the middle of her shift, I was finally able to persuade Jane we needed Hospice staff for their knowledge of pain control. My best selling point was my own tender and subtle experience with Hospice nursing.

As a mother Jeanne was still trying to persuade Jane to eat something almost every hour, repeatedly telling her skeletal daughter, "You are getting way too thin."

Folding laundry with Jeanne in the basement, I tried to explain. "Jane's loss of appetite is a normal part of the process of dying, Jeanne. I think it would be less stressful for both of you if you could allow her to decide when she wants something to eat or drink."

Looking stunned, Jeanne turned to me and said, "You don't think she is going to die do you?"

Only Tom, Dr. S, and I knew the extent of the disease in Jane's body. She hid it well. When asked about how she was feeling, she almost always responded with "Fine," a single word that smoothed and soothed, while protecting Jane from the untactful comments of others.

The most important lesson Jane taught me was that we shouldn't try to carry our own burdens. If we can carry the burdens of someone else, it is a gift. Jane taught me that to take over the grief, fear, and anxiety of another is a healing act of sublimation. Only by exchanging our own grief for another's can we find a realistic antidote to our own despair. But first we must allow ourselves to be helped—and sometimes being helped is harder than carrying the fear of another. Jane gave the weight of her worries to me when she no longer could carry them herself.

As the doula for Jane's death, after tucking Eben into his bed, I would drive across Boulder to spend the night at her bedside. Early each morning, I would leave Jane in order to be with Eben when he awoke. Bob and my father bore the brunt of my new responsibilities. Guilt followed me everywhere, as if I was being haunted by my polygamous Mormon ancestors.

In a small crevice between denial and acceptance, just before Jane agreed to help from Hospice, I sought relief for her from Dr. S. The accumulation of fluid in Jane's abdominal cavity was pushing on her lungs, making her unnecessarily uncomfortable and short of breath. I called Dr. S's nurse requesting a house call to remove the fluid. I had experienced profound relief from having my own ascites reduced in this way. Dr. S agreed to perform the procedure, but only if we brought Jane to his office.

I felt furious with him. He must have known, given his familiarity with the extent of Jane's disease, how painful it would have been to move her, even across the room. I started with pleasant begging and ended up making angry demands. All I got was the pumping equipment.

I had to recruit another doctor—one licensed and experienced enough to insert the needle of the pump into her abdomen without puncturing a vessel or organ. Dr. Luce agreed to make a house call. In less than an hour, he pumped out four liters of fluid. I will always be grateful to Dr. Luce for that moment of merciful old-fashioned doctoring.

Before Jane died, I asked Tom to give his approval of a burial plot in Mountain View Memorial Park, a cemetery within walking distance of the Help's home. My hope was that proximity would facilitate the mourning of Jane's children in the same way Bob had gifted me with a neighborhood burial plot in Columbia Pioneer Cemetery. Afterwards, I went alone to make advance arrangements at the nearest mortuary.

Knowing the Helps had no funds for a funeral and burial, I needed an estimate. Jim Murphy, a young mortician then managing Crist Mortuary, instantly erased all previous impressions of the profession.

I started to provide him with a brief description of the situation, "A young mother with eight children. . ."

Before I even finished, Jim Murphy took the weight from my shoulders, "Don't worry," he said. "There will be no charge. We will do everything we can to assist the family."

Jane asked that there be no visitors as she began her disengagement. Incapable of saying formal goodbyes to her children, I watched as Jane gave them the gift of letting them mother her with tender looks, soothing touch, a teaspoon of sherbet, an ice chip, lotion on her feet, leg rubs, and pillow adjustments. The children and Tom surrounded Jane's bed, filling her last days with moments so tender they were unbearable for me to witness.

In her final days, Jane slept more, but she heard all the sounds—the stories Tom told me of their courtship, the school reports, the discussions about dinner, the frantic searches for shoes and gloves as the car pool waited, and the always alert Pulik parents announcing an unknown human was approaching. Nearly unconscious most hours, whenever Jane stirred, all the children, as well as Duncan and Denevere, would come to her bed, anxious to hear her voice.

On Friday night, November 20, 1992, Jane awoke in the middle of the night and began what sounded like an oral life review, one full of humorous hallucinations. She relived her birthing experiences, talked to babies, asked that the Cool Whip be passed, made endless grocery lists, and tasted the feet of the toy moose she bought to honor "Morty," the star on her favorite television show, *Northern Exposure.* Declaring Morty's feet inedible, Jane drifted into sleep holding him tightly in her arms—an end-of-life lovey.

At the sound of her voice, Duncan and Denevere leapt joyfully onto her bed. After Jane drifted away again, her canine companions were gently lifted off her bed to whimper beneath for hours.

On Saturday night, Jane had another period of alertness when Katherine returned home and began animatedly reporting on her ice skating party. As Katherine spoke, Jane opened her eyes and focused on the crowd in the room. She then greeted each of her children by name.

Towards midnight Jane awakened again and said, "I love you. I am proud of you."

On Sunday, November 22, 1992, Jane was quiet all day and into the evening. Before going to bed, Addie and Katherine took brief turns lying next to their mother. Adam fell asleep on the couch. When I arrived to accompany Jane through another night, the children reluctantly went to bed, as did Tom, who carried his slumbering five-year-old son down to the waterbed.

I was sitting by Jane's bed, holding her left hand, but I must have dozed off because Duncan put his warm snout on my thigh to tell me Jane was gone. A slow pulse had been visible in Jane's throat when I arrived, but it was hard for me to fully focus in the dim light, so I listened carefully to the low hiss of the oxygen tank. I waited for a raspy response from my dear friend's lungs. It did not come.

I let go of Jane's hand and reached toward her neck to check her left carotid for a pulse. As I did so, a strand of vapor moved slowly upwards from her skeletal body. When the twisting vapor disappeared into the low spackled ceiling, I put my face on Jane's abdomen, smothering a long whimpering wail in her stale bedding. It was over. Jane Help died just before two o'clock in the morning, November 23, 1992—the same day she had given birth to her daughter Katherine twelve years earlier.

I crept down the stairs and called Tom's name softly. He was snoring, but seemed to startle awake from apnea and saw me standing in the doorway.

I said, "She's gone."

Still in his street clothes, Tom wordlessly climbed to his wife's side. There was a box of half used pillar candles on a shelf in the laundry room, I waited several minutes before taking the box upstairs and placing the candles on the window sill behind Jane's head. The candlelight, combined with the reflected light from the snow on the ground, created a luminous halo around her body.

The children were each gently awakened, told, and held tightly. Three family circles were held around the bed, the children taking both of their mother's hands in their own to complete the circle, as they said goodbye. Tears flowed, images danced in the candlelit window from Jane's palpable presence. The older girls washed their mother's face, hands, and feet with warm rosemary-scented wash cloths, as her younger children buried their tear stained faces into cloths of the same scent. Questions were asked and answered; embraces substituted for overwhelming pain and loneliness.

During the last hours of darkness, Jane's mouth slowly closed, changing the expression on her face from a gaping open mouth, formed by her long struggle to breathe, to a peaceful subtle smile. Before the children reawakened, the Hospice nurse arrived to certify the death, followed by the Crist Mortuary staff.

When I returned in the afternoon, I found Katherine, Addie and Adam on their mother's empty bed—a snuggery trio in search of some semblance of remembered warmth.

-28-

A Mother's Funeral

Jim Murphy kept his generous promise. Crist Mortuary provided a casket, arranged for a viewing period, provided all the necessary transportation, and handled the burial at Mountain View Cemetery. At the mortuary, I dressed my dear friend's body in clothes carefully chosen by her daughters. At the insistence of the children, Morty was added to their mother's casket and Tara placed a bracelet she had made on her mother's wrist.

A British-born friend once told me she believed the central purpose of the Church of England was to provide a place on three different occasions: at birth, at marriage, and at death. Or as Roland Evans confirmed, "hatch, match, and dispatch." Mortuary chapels sometimes provide a place to dispatch secularized Americans, but the chapel at Crist Mortuary was very small. As a Le Leche leader, Jane had provided support to many mothers in Boulder County; the Friends' School families would be a large crowd, not to mention the friends and families of the older children, as well as staff and clients of Tom's business.

Before Nuhiela's death, Margaret Clifford, a member of St. John's Episcopal Church, had invited Jane, Nuhiela, and me to attend a private healing service there; Jane and I had both wept through those private prayers. Since St. John's had provided Jane comfort while she was living, it seemed the best choice. Margaret obtained permission from Father Rol Hoverstock, who asked to meet with Jane's children.

Extremely kind and gentle, Father Rol asked, "What are your questions?"

Jane's children, all untouched by formal religion, sat stiff and mute in his office, so Father Rol quickly changed tactics: "Episcopalians believe that there is an afterlife—and that death is not something to fear, but to celebrate. For this reason," Father Rol continued, "I will be wearing white vestments in celebration of your mother's life. I want you to feel free to come and see me if you have questions or need to talk over a concern."

Neither Tom, nor any of his now-motherless children, had the requisite wardrobe for public mourning in an Episcopalian church. There was a scramble to assemble appropriate outfits, a task that provided a modicum of distraction.

Jeffy Griffin helped Katherine, Addie, and Adam weave a ceremonial basket to be filled with the sprigs of rosemary we planned to give mourners as they entered the Chapel. I had been boiling sprigs of rosemary on the stove for weeks, using it to scent the wash cloths that had soothed Jane and the tear stained faces of her progeny. Aunt Mary's remedy for tears and the herb of remembrance now belonged to Jane's children.

Crist Mortuary sent the traditional black limousine. Because Tom had asked me to ride with the family, Bob drove me across town to honor this request. Tom also asked me to give the eulogy, a task so challenging that I had slept only a few hours. I was wearing a ridiculous number of hats—the weight of them surreally stacked upon my head.

The funeral was held on Wednesday, the day before Thanksgiving. As expected, an overflow crowd filled the large main chapel. Jane's casket, covered with a white cloth, was wheeled down the center aisle followed by Father Rol in his white vestments singing a psalm. Delivering the eulogy, I attempted to answer the questions I imagined were in the minds of most mourners, while simultaneously trying to sooth the pain of Jane's progeny.

The Help children slowly followed the casket of their mother out of the dark chapel into the bright sunlight. As soon as the doors of the hearse closed on the coffin, mourners crowded around the children and Tom, a landslide of affection and sympathy flowed over them. The handmade basket was chaotically filled with the crushed rosemary sprigs, now scented with the cellular oil of the mourning multitude. Katherine tied a bouquet of helium balloons to the basket and we watched as the basket slowly rose and disappeared into a bright blue Colorado sky.

St. John's Altar Guild, including Martha Hoover, hosted the mourners afterwards in the reception hall. As Father Rol had announced, the burial was private. Following the reception the Help family was taken by limousine to Mountain View Memorial Park, where they awkwardly surrounded the grave as the casket was lowered into the earth. Dismissing the limousine, Jane's eight children and widower husband slowly walked home to Duncan and Denevere.

Thirteen years later, when Katherine and Addie were twenty-five and twenty-three years old, they had the same singular memory of their mother's death and funeral. Each only remembered that after services ended, they had made their way with Adam into St. John's basement classrooms to play hide and seek with Eben and other children from Friends' School.

Hide and seek seemed a fitting denouement to their extraordinary ordeal. It was also the first moment after Jane's death that her three youngest children felt able to return to childhood—wounded, but desperate for a more *optimal distance* from death.

– BOOK IV –

MY BOOK of RUTH

"Where you go, I will go, and where you stay, I will stay. Your people will be my people and your God my God. Where you die, I will die, and there I will be buried."

Book of Ruth 1:16

-29-

Karma in a South Florida Shtetl

Ruth Cantor Pelcyger, Brooklyn, New York, November 15, 1936

Six months after the death of Jane Help, my youthful inspiration to become a heroic doctor in Africa boomeranged. Struck by my own karma, I instead became a self-sacrificing medic for my seventy-nine-year-old mother-in-law, Ruth, in her closely guarded condominium complex in Hollywood, Florida.

In 1989, when Ruth was told that I was now a Hospice patient because I had failed chemotherapy, Ruth's response was to tempt fate.

"Thank God, I've been healthy all my life!" Ruth declared, punctuating her gratitude by knocking her knuckles against an oak cabinet in my kitchen.

Stung, I opened the refrigerator door to hide my tears. Ruth was fresh from Florida. On that day in November 1989, our similarities were limited to our height and our profound love for her first-born son.

Ruth Cantor Pelcyger and her first-born son, Robert Stuart, Brooklyn, 1944

Ruth was then seventy-five; I was forty-seven. Ruth's short grey hair had been professionally styled with permanent products; my own hair was missing. The skin on Ruth's face was invisible beneath a thick coat of foundation; my own skin was peeling and ghostly. Ruth's eyelids and lashes drooped under an excess of blue mascara and eye shadow; my eyelids also drooped even though they lacked lashes. Her expensive two-piece dress, was accessorized by three gold necklaces, an equal number of rings, a gold watch, and large jeweled earrings. By comparison, I resembled an abandoned manikin.

My relationship with Ruth had always been cautious, never close. I was one of the few Gentiles in her life. Ruth was never deliberately mean, just inadvertently tactless—unable to hide the hostility that comes with tribal differences. The first time Bob took Olivia and me to visit Ruth and Eugene it was for Thanksgiving at their Valley Stream, Long Island home. I brought her a large bouquet of beautiful fall flowers.

As I handed the carefully transported orange and red stocks to her, Ruth prophetically responded, "I don't like fresh flowers—they always die!"

The first thing I saw when I entered her Valley Stream, Long Island home was a large couch completely obscured by a custom-made protective plastic cover. Thinking of Ben Evans, I looked at Eugene to see if he appeared nap-deprived.

During dinner, I asked Ruth about the spices she had used on the turkey to make it taste so delicious.

She answered, "Maybe I'll tell you sometime in the future—if you and Bob are still together!" Then she reached across my plate to butter Bob's bread.

Ruth's tricky interrogatories often made me think it was she, rather than Bob or Eugene, who should have become a lawyer.

Her specialty was gift torture: "How many times have the girls worn the muumuus I sent them from Hawaii?"

Answer, exactly once, for less than two minutes, while reluctantly posing for a thank you photograph.

Alternatively, "Did you pick this carpet out by yourself or did you have help?"

Did she like the carpet or should I quickly hire an interior designer?

Two years after Ruth superstitiously knocked on my kitchen cupboards, cancer knocked back. Following breast and lymph surgeries, Ruth underwent six months of chemotherapy treatment. Less than a year later, cells from her breast cancer spread to her brain. After surgery to remove the brain tumor, she endured six weeks of full brain radiation. Sadly, Ruth experienced only a month-long reprieve before the brain tumor "progressed."

On the day Ruth's oncologist told her he had nothing more for her in his doctor's bag, she telephoned us, feeling understandably angry, betrayed, and confounded by modern medicine.

Weeping through her words, she said, "I don't know what to do!"

A long unbearable silence followed.

Desperate to block our collective fall into the chasm of cancer, I blurted, "If you want to try Dr. Gonzalez's treatment, I will meet you in New York and go back to Florida with you to help you start the program."

I made this offer to help, even though I didn't want to spend another minute with Ruth of whatever time remained of my life—a feeling I am certain was mutual. While I was being treated by Dr. Gonzalez, Ruth had repeatedly resolved she could never do what was required of his patients. However, anyone driving through a lethal storm is capable of making a sudden last-minute lane change.

My offer to help Ruth was equally impulsive in the way low heroism makes us blind to danger, drawing us into wars and other risky roles. Maybe I also offered to join the battle against Ruth's brain tumor to help my husband by saving his mother. Albeit, I may have unconsciously been counting on Bob to save me, to refuse to allow me to enlist in his mother's service by giving me a medical deferment.

Bob didn't say, as he once had about my Fort Wayne consulting assignment, "You can't do this! You have to stay home and do your enemas!"

But what man would volunteer to make a *Sophie's Choice*[42] between his wife and his mother? As soon I made my impulsive offer, I began hoping that Ruth would turn me down. But like most of us, Ruth didn't want to die. For the first time in our twenty-year relationship, there was something Ruth wanted and needed from me.

At the karmic moment I offered to help Ruth in April 1993, it had been a year since my radical remission—three years since I had become a patient of Dr. Gonzalez. I was one of a very small group of people (.001 percent) who had survived, not only a six-month prognosis handed down by my oncologist after cells from breast tumors became liver metastases, but also the three-month death sentence pronounced by the neurosurgeon after the cancer had spread to my brain. I was supposed to be dead nine hundred ninety times over, yet I did not believe that Dr. Gonzalez, or any of the other treatment remedies, had permanently cured me. Remission in metastatic cancer is not a full pardon; it is only a temporary reprieve.

From the moment Dr. Luce had called on Passover with the unbelievable news that the MRIs of my brain and liver were clear, I had found it extremely difficult to be a role model for Dr. Gonzalez's metabolic therapy. Then, preoccupied by all that I had uncovered about my own history in the months preceding, I decided that my remission was one of many statistical anomalies found in medical journals, not living proof of the efficacy of Dr. Gonzalez's work. I often felt as if I was back in Utah where I had to pretend to believe in order to survive. Many days, I assumed that the radiologist was incompetent or had mixed up my scans with those of another patient.[43]

I also didn't want my remission labeled miraculous. A search of cancer literature provides evidence that in every type of cancer, as well as in all cancer treatments, there have always been a few inexplicable and remarkable radical remissions. Nevertheless, I did try to make sense of the turn in my fortune.

At my first appointment after the remission, I asked Dr. Gonzalez this question, "When you tell a patient they can get well if they want to, are you using such messages as a strategic aspect of your approach to treatment?"

Dr. Gonzalez's immediate response was, "In a few short years, it will be considered malpractice for any physician to tell a patient that they have only a few months to live."

Pushing further, I asked, "Is your metabolic therapy a placebo in disguise? Are you really practicing mind-body medicine?"

Dr. Gonzalez didn't answer this question; instead he immediately veered away from the subject. However, mind-body medicine is what I have come to believe Dr. Gonzalez must have practiced on me, either deliberately or inadvertently. I also believe that my daily practice of hypnosis acted to potentiate whatever activated my immune system. This is not to negate the obvious impacts Dr. Gonzalez's methodologies have on the body of a patient. The impacts can be beneficial, but can also produce potentially lethal side-effects. Metabolic therapy has real risks.

But back to Ruth. Becoming my mother-in-law's caregiver quickly became a triathlon test of my character. The role caused my normally suppressed cynicism to grow faster than the most aggressive tumor, while simultaneously bringing me close to collapse from a broken heart. Throughout my experience I repeatedly thought about the Biblical story, The Book of Ruth, which is the story of a Gentile woman, a Moabite, named Ruth, who marries a Jew named Boaz, the son of Elimelech and Naomi. In most biblical literature anyone who is a Moab is associated with hostility to Israel. After Boaz dies, Ruth continues to regard herself as a member of his family. She accepts the God of the Israelites as her God and the Israelite people as her own.

In the story, Ruth tells her Israelite mother-in-law, Naomi, "Where you go, I will go and where you stay, I will stay. Your people will be my people and your God my God. Where you die, I will die, and there I will be buried."

That Ruth carried the name of the most famous devoted Gentile in the Hebrew Bible was only one of many ironies in our relationship. That she was Jewish and ended up being dependent on me, her Gentile daughter-in-law, while both of us were suffering from twin cancers is another. A third irony is that as I packed my bag to meet Ruth in New York for her initial consult with Dr. Gonzalez, I did so believing that I would only be away from my son and husband for about a week. Proof that despite my remission, I was still suffering from incurable naiveté.

What follows is my own non-Biblical version of the "Book of Ruth"—all are excerpts taken directly from my 1993-1994 diary entries.

-30-

Less Than Optimal Distance

<u>Day Ten:</u> It has been seven days since I moved into Ruth's South Florida condominium; ten days since I met Ruth in New York City for her appointment with Dr. Gonzalez. Ruth's condominium is one of hundreds in a large Hollywood complex called Hillcrest. The units are occupied by several thousand people, all but two of whom are Jewish—a modern day *shtetl.*

From the floor of Ruth's guest bathroom, where I am now carefully retaining my third coffee enema (an almost unspeakable aspect of Dr. Gonzalez's treatment), I am surrounded by hundreds of blue and white striped cats, their yellow eyes peeking out at me from between silver-foil foliage of the wallpaper. The first time I saw this wallpaper, I felt almost giddy. It made me feel more powerful, more confident of myself because I knew Ruth would never have chosen me to be her first born son's second wife. So I used what I judged was Ruth's extremely poor taste in wallpaper as a social antidote to feeling less than genetically acceptable.

Ruth Pelcyger standing in the doorway between her kitchen and dining room in her Hillcrest Condominium, Hollywood, Florida 1990

I am struggling to find ways to survive in Florida's summer heat and humidity. Ruth is ill enough to be cold all the time, most comfortable when the thermostat is set for eighty five. When she is sleeping, I quietly tiptoe to her refrigerator, open the door, and let the expensive chilled air flow over my body.

Ruth's hearing reflects her age. The sound of television is a comforting companionable white noise, so she keeps her set on at a high volume. She assigned me to sleep in the den next to her bedroom, but even with the den door shut, I can hear every word of dialogue and advertisement with high resolution. For the past two nights, after Ruth has fallen asleep, I have turned down the volume. Both times she awakened, instructing me in a firm, icy voice to turn the volume back up.

We are a distorted version of the odd couple. She likes warmth; I prefer coolness. She needs noise; I am desperate for quiet. Ruth's idea of a good day is to receive a dozen phone calls of short duration from friends, none of which are too inquisitive or revealing. My own satisfactions come from silence and true intimacy.

I do things she would prefer that I not do, like walking barefoot on her carpet and drinking tea. She finds the tea only slightly less dangerous to her carpet than my bare feet.

She has repeatedly asked me not to sit on her white damask living room couch, saying, "I worry that you will stain it."

In response to this controlling insult, I make an effort to sit on it whenever she is sleeping.

Even keys are an issue. Ruth is taking the drug Dilantin to reduce her seizure risks so she can no longer drive. I am no longer taking Dilantin, so I am her chauffeur. However, I am not permitted to keep the keys to her car in my purse, nor have I yet been allowed to drive or walk anywhere alone.

Ruth hasn't even been able to entrust me with a key to her condominium. This means I must plan my frequent trips to and from the fifth floor laundry room very carefully. If she falls asleep while I'm gone, I have to wait in the hall like a suspicious transient. Yesterday I was locked out for two hours.

Fear is the pervasive emotion at Hillcrest—like living inside my mother's paranoia. Android-like security guards patrol the parking lots in golf carts, following residents from their cars to the front door, while another guard watches them as they move through the building on four different monitors. Even the lizards are suspiciously skittish.

<u>Day Eleven</u>: I'm on my back on the floor of the guest bathroom counting cats in a futile effort to lure my psyche into another dimension. I can hear Ruth on the telephone with her best friend Lauretta reporting on the rigors of Dr. Gonzalez's treatment protocols, providing a detailed log of who called today, and reminding Lauretta to acquire an anniversary card for mutual friends.

A few hours ago, while trying to make room in Ruth's kitchen cupboard for the numerous bottles of Gonzalez supplements, I accidentally broke a three-section porcelain platter with a central dipping cup. I didn't tell Ruth about the breakage because I don't want to have to endure her verbal abuse about my carelessness until one of us dies.

As the platter shattered on the kitchen floor, I told myself that Ruth wouldn't be around to look for it the next time it would be needed. I believe that will be after her funeral services at the Levitt-Weinstein Mortuary Memorial Chapel, when her family and friends will come to sit *Shiva* in her condominium.

If my prediction is wrong, and Ruth recovers enough to stage a big party for which she needs the three-section porcelain platter, my whole life will be changed. Not only will I be buying Ruth a replacement, I will also be transformed into a dedicated roving missionary for Dr. Gonzalez and his metabolic cancer treatment. But for now I am trying to hide both my carelessness and my apostasy.

<u>Day Twelve:</u> It is Memorial Day. I was awakened by Olive E. Waite, a part-time health aide, who has been working for Ruth since her diagnosis. Tall and solidly stout, Olive is a fifty-eight-year-old black woman born in Jamaica. Olive has a very close relationship to her God and manifests the confidence that comes with absoluteness in matters of faith. The mother of eight just-grown children, Olive arrived at seven o'clock this morning, having already given her husband an enema. A devout Seventh Day Adventist, she launched into Ruth's enemas with extraordinary enthusiasm.

Snapping on a pair of disposal latex gloves, Olive went right to work. "Ok, Ruthie, now pinch your buttocks together! No, Ruthie! Don't let go! No, No, Ruthie! Don't step in that now. Just let it go."

Olive is a fearless caregiver! Watching her, I realized I had approached Ruth's enemas with timidity and too much distance. Personally, I would have quit the Gonzalez program immediately had I needed any kind of hands-on assistance. I was in naive denial when I offered to come to help Ruth since it had never occurred to me that it would be necessary for me to provide such intimate support. At one point, Olive stood, holding the enema bag in the air above Ruth, who was laying on the bathroom floor, surrounded by berms of soaked towels, as I attempted to hold the tube in Ruth's anus with my left hand, while wiping up the worst part of the mess with my right. The disease has suffocated her normal sense of modesty. She makes no protest.

They don't make movies with these kinds of scenes because, as Eben described enemas when he was seven, "They aren't that interesting."

Today is not only Memorial Day; it is also Maxwell Yellen's eighty-second birthday. Four feet, ten inches tall, with still-thick white hair and even thicker eyeglasses, Maxwell is four inches shorter than Ruth. He has been her romantic companion for decade. Ruth's husband Eugene died of a heart attack in 1979. Maxwell and Ruth are also astrologically close since Ruth's seventy-ninth birthday is tomorrow. All three of us were born under the sign of Gemini, but not one of us believes in astrology.

Ruth invited Maxwell to go out for his birthday dinner. In my new role as Ruth's medical coach, nurse, dietician, cook, chauffeur, laundress, and personal shopper, it was necessary for me to be the third wheel on their dinner date at the Unicorn Village, a nearby restaurant-grocery complex, offering sustenance that supposedly resembles Dr. Gonzalez's prescribed organic diet.

As we dined, I asked Maxwell about his life story. The youngest of five children born in Brest, Poland, Maxwell arrived in America in 1914. His father was the first to emigrate, working days as a presser and nights as a watchman in New York City to save enough money for his wife and children to join him in America. Maxwell broke all the rules of his old *shtetl*—along with his mother's heart—when he married an Irish Catholic woman he met in a restaurant on Forty-Second Street. He and his Catholic wife had five children during twelve years of marriage in the Bronx. By my math, Maxwell has been divorced and single for forty-nine years.

What Maxwell didn't tell me, but others had, is that he is living at the edge of financial and social peril. Like me, Maxwell is something of an imposter. He has a fictional real estate practice he pretends to operate out of an illegally-rented condo on the third floor of Ruth's building. Legally blind from macular degeneration, Maxwell has no driver's license, but nonetheless continues to drive his uninsured 1978 Cadillac. When I told Maxwell that Ruth's two sons were very concerned that their mother was sometimes being driven by an unlicensed driver with impaired vision in an uninsured vehicle, Maxwell responded by assuring me that he was a very good driver and would still be driving when he was one hundred.

"Sure" I teased him, "and tomorrow I'm going to be crowned Miss America!"

He didn't laugh. Like others their age, Maxwell and Ruth see nothing coming tomorrow, except more of today.

Day Thirteen: Two weeks ago, I found Ruth waiting for me in a wheelchair in the baggage claim area of LaGuardia Airport. Five months had passed since Ruth had spent the winter holidays with us in Boulder—I was unprepared for the degree of deterioration. We made our way slowly to the line for taxis—Ruth's heavy handbag hanging from my left arm, its owner hanging from my right.

We checked into the Murray Hill East Hotel on Thirty-Ninth Street between Lexington and Third Avenue, a modest establishment with kitchenettes offering a small discount to Gonzalez patients. Struggling like strangers to make ourselves comfortable in a one-bedroom suite, Ruth took the bedroom, while I took the sleeper sofa, and we shared the small bathroom. The next morning, Ruth discovered that she was a woman in New York whose make-up kit had mysteriously gone missing, while I discovered I was a *shiksa* without the requisite lending supplies.

That first day in Manhattan, Ruth asked me more than fifteen times if she could borrow my foundation and rouge. I carefully explained, an equal number of times that I didn't wear these items. I didn't say that I didn't need them, nor did I explain that having grown up Mormon in Utah, something in my psyche still considered make-up to be a kind of sexual contraband. Instead, I offered to take Ruth to Saks Fifth Avenue or a nearby drug store to buy replacements.

"No" Ruth demurred. "They would be too expensive! Lauretta takes me to a *schmatte*[44] row, where I buy them discount."

Yesterday, a full two weeks after the start of the "Mystery of the Missing Make-Up Case" (and after at least two hundred requests to borrow my nonexistent beauty products), Ruth, Lauretta, and Lauretta's friend, Edith (of whom Ruth is "not too fond") finally went to their favorite *schmatte* row for the necessary replacements. Since it was her birthday, Ruth also bought herself a handbag, a three hundred seventy dollar model on sale for two hundred dollars.

Hand-painted in four colors, Ruth handed her new bag to me to inspect, explaining that "It will go with everything."

Ruth needed a nap after shopping and disappeared into her bedroom. Having been involuntarily designated substitute hostess, I served Lauretta and Edith glasses of organic apple-raspberry juice.

Lauretta gossiped about how surprised she was that, "Ruthie had spent that much money on a bag now."

Lauretta also used Ruth's nap time to slice and dice Maxwell's persona. Lauretta's judgment is that Maxwell should be eliminated from Ruth's life because he is a sexual and social dud and, worse yet, a financial liability. A couple of years ago, Maxwell asked Ruth for a ten thousand dollar loan. Ruth gave him the money, despite Lauretta's caution.

Nap over, Lauretta insisted on taking us out to dinner to celebrate Ruth's birthday. While Lauretta applied Ruth's new make-up, Ruth asked me to transfer the contents of her old handbag into her new birthday bag.

Using a heavy hand, Lauretta started by smearing cover-up stick on Ruth's upper lip and beneath her eyes, before covering her whole face and neck with a solid coat of thick liquid foundation. Then Lauretta added eyebrows with a pencil, applied shimmering blue eye shadow on Ruth's lids and dark blue mascara on the stubby regrowth of her lashes. Next, using a red lip pencil, Lauretta outlined a mouth in a shape that bore more resemblance to Lauretta's than Ruth's. After she had filled in Ruth's new mouth with a thick coat of pink lipstick, Lauretta reached for the rouge. It was missing! Where was it? The rouge was the most critical item! I kept silent, but perversely saw it as beneficial coincidence.

Like Maxwell's, Ruth's birthday dinner took place at organic central, the Unicorn Village. There was no birthday dessert with a candle lit by a singing waiter because Lauretta and Edith were already late for a standing game of Mahjongg. Also, there was nothing on the dessert menu that Ruth was allowed to eat. I drove my sulky and petulant patient home.

There had been no call, card, or gift from Maxwell. His previous presents, prominently displayed, were momentarily transformed into painful reminders—the yellow teddy bear sat in its honored place on Ruth's bed; next to a strange phallic-shaped octopus; and on the living room carpet a battery-operated dog waited to be wound up so it could walk while barking at a pitch two octaves above a miniature poodle.

I helped Ruth with her enemas and into her pajamas. Then, while she received birthday telephone greetings from her sons in Colorado and California, I gave her feet an oil massage—something she initially resisted, but now requests. Each son dutifully performed the ritual Ruth has instilled in them by singing off-key renditions of "Happy Birthday" into their far off mouthpieces. Ruth's birthday came to a close when she fell asleep during the local midnight news.

That was yesterday. This afternoon after taking the first dose of Epsom salts for Dr. Gonzalez's "Liver Flush," Ruth insisted that Olive drive her to *schmatte* row for another replacement rouge. Fortunately, the Epsom salts, which act as a harsh cramping laxative, didn't start to work until well after the third dose. I teased Ruth about this, telling her that she would win the gold medal in any Epsom salts competition. When the salts finally took effect, she got the joke.

Ruth is mad at Maxwell. On her birthday twenty friends and family telephoned and, by her careful repetitive count, she received more than fifty birthday cards, but Maxwell made no contact. This confirms Lauretta's judgment that Maxwell is a social dud. His silence surprised even me, but perhaps he feels inhibited by my presence. Or maybe Maxwell is so impoverished a card might break his budget.

As Maxwell accepted Ruth's birthday card in the Unicorn Village, I heard him say in a barely audible voice, "I looked for one for you, but couldn't find one that was right."

Over the years, Ruth's building, and the adjoining Hillcrest structures, have slowly been transformed into a great ghetto of loneliness. Many units are empty because Hillcrest covenants prohibit renting. Others are dark and quiet because their Snowbird occupants have gone north for the summer. Maxwell is an archetype of social isolation. His parents and siblings are dead; he is long divorced, perhaps abandoned by his children and barely surviving on Social Security. Maxwell is only a millisecond away from disaster—a missed step, a missed heartbeat, a missed stoplight, a missed toilet stop, or a missed birthday card. A missed call might be the first sign that he has died, his body silently decaying one floor above us.

In a futile attempt to distract Ruth from her Hallmark disappointment, I asked about her childhood. Ruth is one of ten children born to an immigrant couple from Minsk. Her father, Samuel Cantor, immigrated to America in 1904. Her mother, Sarah Rebecca Bindler, followed a year later with their first three children. Seven more were born in America, of which Ruth was the second to last.

The couple settled in the Brownsville area of Brooklyn as part of a close-knit group of immigrants from Minsk and nearby *shtetlekh*. Ruth's mother, Sarah, was an exceptionally loving soul, who died of cancer before Ruth's marriage to Eugene. Samuel died many years later, after having remarried her mother's nurse and housekeeper; the latter was an act for which he was only partially forgiven. Ruth's well-preserved resentment of her father's remarriage made me wonder how Bob and Joel would respond if she re-married Maxwell.

Like Bob and me, Ruth and Eugene met at work. Ruth was working as a legal secretary in the law office of Arthur Goldstein, where Eugene was a young associate attorney. In romance, proximity at work often defines one's destiny.

<u>Day Fourteen</u>: As a foreigner in Florida, it is necessary for me to pay close attention to my surroundings. Florida is a place where African violets flourish, but the population prefers the artificial version. It is a place of overly-penciled lips, of pink-silver hair, of sweatshirts decorated as if they were evening gowns, and of oversized cars, which, when viewed from the rear, appear to be moving without drivers.

Florida is a place, where, when ordering an "Early Bird Special" in a restaurant, one eats only enough to leave some to carry home. I doubt that much of what these Early Birds carry home is ever eaten. By my sixth day here, Ruth had acquired three bags of Early Bird leftovers, but eaten none—even though she sent me out at midnight in a torrential downpour to retrieve an Early Bird bag from her car, where it had been mistakenly stored at more than ninety-five degrees for six hours.

I certainly wouldn't touch the contents, but she won't let me throw the bag away. Ruth has a brain tumor that has all but obliterated her short-term memory, yet she remembered the remains of her Early Bird Special.

-31-

Preparations for Both Outcomes

Day Fifteen: This morning I raised the issue of Hospice. Olive had more trouble with the topic than Ruth.

Wanting to talk in private, Olive whispered, "Did the doctor tell you something that Ruthie don't know?"

Explaining my thinking, I said, "No, Olive, but we have to plan against both outcomes. Hospice can be incredibly helpful in end-stage pain management. On the other hand, if Dr. Gonzalez's therapy works, Hospice staff will be very happy to see Ruth recover."

Olive is afraid that she won't be needed if Hospice shows up. Speaking without authority for Ruth's two sons, I reassured Olive that we needed as many hours of her time as she was willing to give. I talked openly with both Ruth and Olive about the two possible outcomes. Ruth could get well or she could die. In either case, Olive eventually would need to find other work. To reduce her concern, I promised Olive sixty days severance pay, a pittance for her enthusiastic enema expertise.

Ruth is in a desperate struggle to know what is coming next and to remain on top of things. She reminds me of myself during chemotherapy—trying to prove I was still who I had once been; striving to always appear as normal as possible in front of Eben, Bob, and others. I drove myself to Denver for treatment, sometimes as erratically as the first time I got behind the wheel of my father's 1939 Chrysler Imperial. Acting as if nothing was wrong inside my body, constantly humming to distract myself from pain, I was desperate to avoid asking for help. It isn't easy to watch. Others instinctively stand back when they see the denial of death in someone's face or sense the raw fear of death.

Ruth's friend, Ina, a former bookkeeper, came by this afternoon. Ruth didn't bother wearing a wig, wandering about distractedly, while Ina balanced Ruth's checkbook. I occupied myself with the endless laundry that is a side effect of incontinent bowels. Between laundry cycles, I threw away stacks of obsolete duplicate Medicare bills, but only when Ruth wasn't looking. Ruth is unable to let go, preserving her lifelong habit of indiscriminate saving, complicated because the capacities of her mind and body are unraveling at warp speed. Ruth needs as much of her life as possible to remain stationary.

Alone in the laundry room, I wept, my tears dampening her sheets. Overwhelmed by fatigue and hopelessness, I want us both to just give up and die.

With respect to Dr. Gonzalez's treatment, Ruth is a moderately regressed patient. She is doing everything he has instructed her to do, but first she tries to negotiate a better deal.

Ruth's negotiating tactics include, "Why almonds? Couldn't I eat cashews? I like them better?" "Can't I divide these pills—take some now and some later?" "Couldn't I just get all these vegetables in a soup?" "Why can't the salad be shredded and dressed with mayonnaise?" "He said that I could have chicken in six months, right?" (Dr. Gonzalez promised no such thing.) "I just need a little something to help get the vegetables down, like pasta or mashed potatoes."

At our new sub-optimal distance, Ruth's rage at having been abandoned and betrayed by her progeny is impossible for me to ignore. Ruth's two sons migrated west, away from their mother and father. Each married outside their tribe, putting unmeasurable distance between themselves and the world of their parents. There are no *mezuzahs* on the doors of her sons; neither goes to temple, nor do Ruth's grandchildren.

Bob divorced the Jewish woman he married at the St. Regis Hotel in New York City, and then added to his betrayal by replacing her with a redheaded *shiksa* and adopting her Gentile daughter.

Joel works with children, but has none of his own. Instead, he married a Gentile divorcee twelve years his senior, who has three adult children from her first marriage.

Bob in particular has frequently failed to send the critical greeting cards on the right days. When Bob's birthday card to his mother arrived a day late yesterday, Ruth made me read it aloud to her and Olive. Ruth's rigorous Hallmark card coda is that any filial laggard who fails to send her a card will receive a call from Ruth, during which she will deliver her disappointment by reading the text of a card sent by a non-laggard. For example, last year, Ruth called Bob on Mother's Day to read him the text of Joel's Mother's Day card, which had arrived on time, after Bob had not sent her a card of his own.

From this too close perspective, I can see what brutal children Bob and Joel have been. Further, to survive, Ruth must allow me, a Gentile, to penetrate the most secret cavities of her body and her daily habits. A jury of Ruth's peers would acquit her were she to come into the den in the middle of the night and smother me. Truthfully, I would be hard-pressed to fight back.

Speaking of fighting back, I am battling loneliness, regret, and guilt for having abandoned my youngest child. On the morning I left Boulder to meet Ruth in New York, I watched Eben as he walked away from me, disappearing through the front door of his school. I began weeping knowing that I might never see him again—just one way terminal illness contaminates ordinary transactions, interfering with our normal denial of death.

My preparations for leaving Boulder were frantic, even though I expected to be gone for only a week. There was a tenth birthday party to plan for Addie Help, now dripping with despair from her mother's death. The irrevocable spring timetable of my garden was already moving forward without me. The ghosts of my Mormon female ancestors were insistently whispering reminders to leave nothing undone—no dirty clothes, no bills, no dust, no crumbs or sticky surfaces.

Aching with anticipated homesickness, I wrote a thesis of instructions for my husband, my father, the house cleaner, and a temporary afternoon nanny, detailing all the tasks they would be covering for me. Days before my departure, Marley, our Hungarian Puli, and Magic, our cockatiel, detected the scent of approaching abandonment and attached their bodies to mine. Marley's breath dampened the heels of my socks, Magic's the shoulder of my shirt, as if both were begging me not to go.

In Florida, I am my sole consoler. Attempting self-absolution by forcing myself to acknowledge the inevitability of having to let go of all my attachments, knowing that Bob and Eben must learn to live without me. The upside of my absence is that father and son, along with canine and cockatiel, are being forced to practice operating as an amputated family unit.

In our telephone talk, the anticipated pain is anthropomorphized into Marley and Magic. Bob reports that Marley is anxiously waiting for me by the front door, refusing offers from alternate walkers. Eben tearfully tells me that Magic is plucking out his feathers to abate his new solitude in my absence. I cannot hold motherless Katherine, Addie, and Adam, or any of their five older siblings, in my mind's eye without weeping. Swallowing the ache of longing and loss, I attempt to escape into a hypnotic state of cool, pain-free mindlessness—a transient treatment transfused into my body by the taped voice of Roland Evans.

Maxwell is still alive. I ran into him by the elevator.

"What is happening with My Angel?" he asked.

I teased he was in trouble because of his silence on Ruth's birthday.

"But," he said, "I thought we had celebrated!"

I tried to explain, "No, Maxwell, Ruth considers our dinner to have been your birthday celebration, not hers. You better think of a way to make amends."

As soon as I returned from the laundry room, the phone rang. It was Maxwell, a social genius in disguise, calling to ask if he could listen to the tape of Ruth's meeting with Dr. Gonzalez. That was all it took for him to work his way back into his Angel's good graces. Icing this cake of social triumph, he offered to help Ruth with the enemas! All three of us had a good laugh about his offer, but Maxwell's devoted demeanor told me his gesture was genuine. Too bad his eyesight is so poor. Finding Ruth's anus would be much harder for Maxwell than following the center line of the road.

Maxwell's romantic attention makes his social disabilities tolerable. Ruth seems to return only enough affection to keep his adoration flowing over her. Maxwell can't see; he can barely hear; he shouldn't drive; he has no money; and he can't get an erection. But, he does look at Ruth with possessive desire. At their stage of life, intent makes up for other deficiencies.

<u>Day Sixteen:</u> When Sheila, the Hospice in-take nurse, arrived this morning, I had to wake Ruth from a very sound nap. Like Ruth, Sheila was born in Brooklyn. Nurse and patient also have the same length and color of acrylic nails in common. Sheila's body seemed to recharge whenever her pager beeped, which was six times during the hour-long intake interview.

Ruth signed the requisite Medicare, supplementary insurance, and transfer of care documents in preparation for tomorrow when another Hospice nurse will come to do a baseline assessment of her body—the condition of her skin, temperature, blood pressure, etc. Sheila showed only peripheral interest in Dr. Gonzalez's treatment program, since to qualify for Hospice a patient need only agree to give up standard treatment, everything except palliative measures. Hospice, guided by the unwillingness of insurance companies to cover the cost of Dr. Gonzalez's therapy, considers it neither standard, nor palliative.

After Sheila told Ruth she had been an oncology nurse, Ruth assumed Sheila's disinterest in Dr. Gonzalez's approach was a sign that Sheila believed only in chemotherapy and radiation. From that moment on, Ruth regarded Sheila as if she were a Gentile, despite their common origins and beauty practices. When Sheila explained all the aspects of palliative care, Ruth barely nodded, repeating twice that she does not want to be resuscitated. She stated that she has a living will, but when I searched later all I found was her sister Selma's.

Tonight Ruth's mood was elevated. Instead of turning on the kitchen television, she was talkative over dinner, telling me stories about her friends. Uncharacteristically, Ruth deliberately let the answering machine take Selma's nightly six o'clock call. I think she is ready to be relieved of this placebo—an unspoken agreement by two sisters to make a nightly check that the other is still alive.

After dinner, I helped Ruth with an enema and into the shower. While I changed her bed linens, she was strong enough to dress herself and apply her own make-up. Then she asked me to escort her to a fifth floor condo where her friend Lilia was hosting a Mahjongg game. I can only imagine the kind of difficulty Ruth's participation creates.

All of her friends are afraid for Ruth and afraid of her. They don't want to face her deterioration—a living reminder that their own vitality is equally vulnerable. Rapt listeners when I explain what they should do if Ruth has a seizure in their presence, they take on the demeanor of serious students, repeating the signs and symptoms of a petite mal, as if they were learning a new language. But their eyes reveal that they really don't want to be alone with Ruth. She is scary now. Ruth feels this too and we have talked about it. One afternoon when no one called for more than three hours, Ruth became angry.

"Most of my friends are already pulling away from me! I want them to stop treating me like I am already dead!" she said.

After Ruth disappeared through Leila's door, I went on an illegal walk across the links of the Hillcrest Country Club golf course. During a visit four years ago, I saw a red fox with two kits near the fifth hole. Then my heart ached thinking of the mother fox trying to help her kits survive in this urban jungle. I looked for foxes again tonight, but saw none.

After two treks around the course, I crouched beneath a Ficus tree as big as our house in Boulder. Not another living soul was visible; the only sounds came from distant car horns and the snake-like hissing of impulse sprinklers. Walking back into Ruth's world, my feet moved slowly, careful to give enough warning to the lizards skittering ahead of me between the thick blades of Bermuda grass.

Day Seventeen: Yesterday was Friday, Shabbat. Late in the afternoon, Ruth began rummaging around in the kitchen for candlesticks and candles, telling me, "I sometimes light the candles on Friday night like my mother did."

Ruth has smothered any aesthetic appeal in her practice of this traditional Jewish ritual. Instead of the table, the candlesticks had to be set out on the kitchen counter, after which a piece of cardboard was placed beneath them. As further protection, the bases of the silver candlesticks were wrapped in aluminum foil to catch any dripping wax.

Since it was *Shabbat,* I relented and prepared illegal pasta with olive oil, tomatoes and basil for Ruth. I chopped raw vegetables, added them to a green salad, and lightly steamed some asparagus. We ate in silence—Ruth watched *Wheel of Fortune*, while I read a letter from a friend. After Ruth continued to sip her carrot juice so slowly she neutered its nutritional potency, I relented a second time, serving her a plate of fresh sliced peaches dribbled with apricot liquor, which I knew was the source of her carotene stalling.

As the sun set, Ruth and I stood side by side, our backs to the refrigerator.

She lit the candles, saying the *Shabbat* prayers into her cupped hands, inviting me to repeat the words after her, which I did. "*Baruch atah Adonai, Eloheinu melech ha'olam . . .*"

When the sun disappeared, I irreverently suggested driving to the beach. After taking off our shoes, we put our feet into the warm salt water, and watched in silence as a yellow moon moved in and out of dark thunder heads at the edge of our visible world.

The *Shabbat* candles were still burning safely upon our return.

-32-

A Feral Cat at the Westminster Dog Show

Day Nineteen: Today Ruth and I took a journey into the future. We planned to stop at the locksmith to copy Ruth's keys so I had my own set, followed by a stop at Future Nails in order to have two of Ruth's missing acrylic nails replaced. Afterwards Lauretta would join us for dinner to dilute our too-close companionship.

The locksmith shop was closed until later in the afternoon, so I drove west toward Future Nails as Ruth instructed, but she quickly became disoriented. An hour later, she agreed we were lost. I suggested stopping to ask for directions or calling Lauretta for help, since Future Nails was somewhere near Lauretta's condominium complex. Ruth resisted, continuing to berate herself. After more wandering, she found a familiar street and from there, Ruth was able to direct me to Lauretta's condo.

As soon as we stepped inside, Lauretta received an emergency call from one of her sons. Cancelling our dinner date as she gathered her things, Lauretta rattled off directions to Future Nails. Back inside the car, Ruth quadrupled her usual number of back seat instructions, as if trying to make up for her previous disorientation—her repetitions clogging my brain as I drove through a dramatic downpour, hurricane-force winds buffeting the car.

Ruth arrived at Future Nails dry. As Ruth's driver, valet, and umbrella holder, I entered Future Nails sodden and sopping wet. The aroma inside Future Nails smelled like a carcinogenic carnival. The air of the single large room was saturated with fumes from acrylic resins, hair spray, and the scents of about fifty heavily-perfumed patrons being attended to by two dozen manicurists and ten hair stylists.

I felt exactly like what I resembled: a wet feral cat accidently chased into the center ring at the Westminster Dog Show in Madison Square Garden! Among the too-obvious differences: my plain, very damp, wrinkled linen smock dress to their two-piece polyester sports outfits decorated with sequins or studded with fake jewels; my plain wedding band and leather-banded watch to their multiple rings with rock-sized gems and jewel encrusted watches, charm bracelets, necklaces and large dangling earrings; my nude face to their camouflaged visages. Finally, there was the contrast between my short, unpainted nails to their long, brightly colored ones, which, while I admit may have some sex appeal, are nonetheless serious obstacles to independent living. Ruth's acrylic nails make it twice as difficult for her to grasp a pill, use a key, write with a pen, but, most importantly at this moment in our shared existence, to wipe herself.

It is now midnight and I am no longer a manicure virgin. Once Ruth's nails were finished, she insisted in a loud voice that I have mine done. I protested, as one hundred heavily disguised eyes focused on me, the half-drowned feral feline. Then I had a small epiphany. Given the amount of control I have assumed over Ruth's body these past weeks, she had the right to some control over mine. So I sat down at Maria's manicure station, and asked for "a light color." In the cacophony of New York and Spanish accents, Maria must have heard me say that I wanted "white."

Maria began with a scolding, "Why are your hands so raw and red?"

I didn't even attempt to explain that my current work required perpetual hand washing and floor scrubbing with Clorox and hot water. By the time Maria had finished lecturing me on the difficulties presented by my "frozen cuticles" and had applied the polish, Ruth was agitated, restless, and ready to go.

As I moved toward the drying station, still squishing out a trail of wet footprints, Ruth commanded me, in her *El Capitan* voice to take her wallet from her purse, reach into another zippered compartment for her keys, open her umbrella, and retrieve the car. By the time Ruth was safely buckled into the passenger seat, I had been drowned again and my first professional manicure was one hundred percent imperfect.

We drove east toward Hollywood on Johnson Street, narrowed to one lane from water pooled on both sides. Even though the locksmith shop was still shuttered tight, Ruth insisted I stop. As I waded through a foot of water to rattle the door, the experience was familiar. Ruth has taught her sons to also not accept any "closed" signs without a struggle.

When we reached the Unicorn Village, Ruth telephoned her sister Selma to see if she and her husband George King could join us for an early dinner. When they declined, Ruth poured out her grief and rage for feeling she had been abandoned by her closest and only living sister as she fork-stabbed a plate of raw vegetables. As soon as we finished our meal, despite Selma and George having declined her dinner invitation, Ruth wanted to drop by their apartment. After wandering around their Aventura apartment complex for an hour, guided by another bout of denied disorientation, we had a painfully short encounter.

The spark that had once lit Selma's beautiful soul was gone, either by stroke, Alzheimer's, or too many days with her husband, George King. A caricature of the hypochondriac, George's status in life seemed to be tied to the quantity and frequency of his doctor appointments.

Eugene and Ruth Pelcyger with Selma and George King, Connecticut, 1939

Also fixated on food, George kept announcing over and over, "You are what you eat!"

This comment was followed by a long list of all the foods he would not ingest. Condiments were given a particularly nasty verbal lashing. Selma spoke only twice: once to show me her "SAS" (special arch support) shoes and once to answer my question about her own lung tumor.

Speaking without affect, Selma said, "It was benign and the doctor dismissed me."

During our ride home, Ruth laughed when I referred to her brother-in-law "King George." She thought it was the perfect descriptor. It seemed to me that Selma's withdrawal from their sisterly relationship was not personal.

"Selma seems almost gone from the world, Ruth," I said. "She is light to the touch. Her spark has burned out, but yours has not, which is why it is so painful."

Reaching across the seat, Ruth squeezed my upper arm and said, "You should have been a therapist."

Possibly this was a rare compliment from Ruth, but statistically unlikely since she and her friends do not believe in talk therapy.

As Ruth unlocked her front door, the air in the condominium was unusually warm, even for Ruth. Before I finished putting our groceries away, Ruth had completely disrobed. She came into the kitchen to announce she was ready for her enemas. I retrieved the aluminum and nylon camp cot purchased at Kmart to use as improvised enema equipment. More comfortable than the floor, the cot made it easier for Ruth to successfully move to the toilet seat.

The first two enemas were successfully administered, held, and released. But Ruth lost control of the third. She sat on the toilet, while I wiped up the mess and hosed down the cot in the shower stall.

"I wish you didn't have to do this!" she said.

"I know, Ruth," I replied, "but I think we are both learning from this experience."

It would be out of character for Ruth to ask exactly what we were learning, but from my perspective the psychological distance between us was shrinking. Despite our fundamental differences, we were becoming closer in our mutual struggle to survive.

Ruth napped until after nine. When she awoke, I made her a late supper of carrot juice, vegetable soup, and a soft-boiled egg, served suffused by an episode of *Nurses* on the kitchen television. There was one enema joke and several about naked buttocks, but neither of us laughed.

After Ruth went to bed, I removed what was left of my white nail polish.

<u>Day Twenty-Two</u>: I have had little time or energy to write. Caregiving has consumed all of both. Even when Olive is present, Ruth is reluctant to let me out of her sight. Olive has tried to run interference for me.

"Now, Ruthie, your daughter, she needs her rest!" Olive says emphatically. "You must let her go by herself! Joanie needs to find her own place here!"

Olive tells me when Ruth is sleeping, "Joanie, you have gone from being a daughter-in-law to a daughter because, believe me, I knows that no daughter-in-law would do this for a mommy-in-law!"

Because Olive works an eight-hour shift on Wednesday, I made arrangements to visit Carrie, the only maternal cousin of Bob's ex-wife, Deborah. Carrie has just been diagnosed with pancreatic cancer. Her request to meet with me an example of how fear of death dissolves the boundary lines created by divorce.

Carrie decided her best hope for surviving pancreatic cancer longer than the twelve-month average was Dr. Gonzalez's metabolic therapy. She had traveled to New York a few weeks before Ruth for a consultation with Dr. Linda Lee Isaacs, Dr. Gonzalez's wife and newly licensed medical partner. Carrie took an appointment with Dr. Isaacs because her estimated lifetime seemed too short to wait for the return of Dr. Gonzalez from a two-week speaking engagement in Europe.

I had glimpsed Dr. Isaacs when I accompanied Ruth for her appointment. Dr. Gonzalez had previously told me they would soon move from their current cramped suite on Park Avenue into a larger suite of custom-designed offices on East Thirty-Sixth Street, across from the Morgan Library.

Ruth was anxious and envious about my going to see Carrie without her. With Olive's support, I remained firm. I did not want to impose Ruth on Carrie, despite the fact that several patients had convinced me to stick with the Gonzalez program at critical junctures. But Ruth is either not far enough along or too far gone to be a hopeful model.

While Ruth and I were in New York, I had a poignant breakfast meeting with Harris Zidel from Garberville. Nonny gave Harris my name after he was diagnosed with pancreatic cancer and decided to seek treatment from Dr. Gonzalez. We met at the Cupping Room Café on Broome Street, not far from the World Trade Center. Ruth had relentlessly interrupted Harris with self-centered one-to-one comparisons.

Harris and his wife Peggy have one child, a daughter named Sophie, the same age as Eben. Harris has a great sense of humor and we had developed an almost daily telephone relationship, trading tips, tricks, and troubles, in between bouts of laughter. After a year of unusual long-distance intimacy, it was lovely to meet and talk in person. It was also inspiring to see a person with pancreatic cancer, mobile and lively. At breakfast, Harris wolfed down a double order of eggs benedict. He seemed a little reckless about Gonzalez protocols, something I understood from my own deviations.

Later, when I mentioned our meeting to Dr. Gonzalez, he said, "Oh, yes, Harris. He needs to do things his own way. It makes me nervous."

I don't believe that there has ever been or ever will be a perfect Gonzalez patient, one who does exactly as instructed one hundred percent of the time. Compliance is a constant challenge. While an oncologist knows exactly how many units of chemotherapy have been administered and a radiologist the number of rads (radiation absorbed doses), Gonzalez patients are solely responsible for their own therapy. He wants us to fully insert the thirty-inch extension that comes with his prescribed enema kit, but enemas are a particularly difficult treatment component to measure—a serious and unpleasant complication in any clinical trial.

At a residential complex in Coconut Grove, I found giant verdigris gates and an armed guard protecting Carrie, her second husband, their twenty-two-year-old son, and fourteen-year-old daughter from the outside world. After the guard received telephone permission from Carrie, the gates opened and I passed into a different economic stratum.

At age fifty-three, Carrie's tall graceful beauty matched that of her elegant three-story home. Carrie had an extremely abrupt introduction to cancer. Unlike most cancer patients, who receive an initial diagnosis, take treatment, and, if lucky, have a period of remission, Carrie was hospitalized with phlebitis and discharged with terminal pancreatic cancer. Her face was a portrait of shock and Carrie's family was also reeling.

After listening intently to Carrie recount her story, I could only offer paltry, cheap advice, and a few technical suggestions. My most useful contribution was to juice carrots in Carrie's immaculate, rarely-used dream kitchen.

Driving back to Ruth, I began imagining the kind of treatment center that should be created for Gonzalez patients. It wouldn't require Doctors Gonzales or Isaacs in person, but rather a crew of loving and fearless souls (preferably all clones of Olive) intimately familiar with all the techniques and side-effects of the various protocols. These nurses and aides would need access to doctors capable of suppressing enough skepticism to make clinical assessments that are so necessary in patients with metastatic disease.

If new patients, along with one or more of their caregivers, could spend even a few days at such a place, the hands-on-expertise of staff would be extremely valuable. For Ruth and Carrie, it could mean the difference between sticking with the program versus dropping out. Both started when they were already too ill to manage without supportive, knowledgeable help. I see the absence of such support as a dangerous gap. My day-dream was conjured up not only by my own experiences, but by my desperate desire to have my current responsibilities outsourced. As a caregiver, I resemble a lonely ember in a dying fire.

-33-

Illness, Intimacy, and Social Abandonment

Day Twenty-Five: It is almost midnight. I have been on the telephone with a Hospice nurse trying to decide whether to repeat Ruth's medications (Dilantin, to prevent seizures, and Decadron, to reduce inflammation) because she vomited after dinner. Since the medications were in Ruth's stomach for more than one hour, we decided not to repeat the dose, but to move up the next one by several hours.

Olive had a dental crisis so I have been on duty for four straight days and nights with no relief. Completely exhausted, my brain is behaving like Ruth's. This afternoon I put her enamel teapot on the stove, then promptly forgot it. Even though I smelled something burning, the alarming scent failed to jog my memory. By the time I returned to the kitchen, the enamel had melted onto the electric unit, permanently sealing teapot to stove.

Ruth slept through most of the day. A grand mal seizure last night left her rag-doll-limp. I was awakened in the first moments of her seizure by the sound of her knocking the lamp off her bedside table. Rushing to her side, I almost lost a finger in my attempt to keep her from biting her tongue. She is stronger than me in many ways.

This morning Ruth had a bowel accident that started in her bedroom on the shag carpet and ended in her master bath. Despite the need for cleanup, she wanted to proceed with her morning enemas. I refused, my patience with my patient spent. After helping Ruth take a shower in the cat bathroom and getting her dressed in clean clothes, I persuaded her to wait in the living room. Ruth napped on her white satin damask couch, while I scrubbed and scoured the accident sites. When I woke Ruth for her eleven o'clock dose of pancreatic enzymes, she was ready to start on her enemas. She immediately began stripping off her clothes, everything except her blouse.

Attempting deterrence, I suggested, "Ruth, we probably shouldn't start your enemas yet. The smell of Clorox in the bathroom is too strong. Let's wait until after lunch."

Addressing me as a simple soldier beneath her rank, *El Capitan* Ruth did not agree.

"No!" she said, "I want to get on with my program!"

My voice, a reflection of my insubordinate morale, was flat, "One missed enema is not a life or death issue."

Ruth responded to my denial of service with body language. She remained half-dressed and splayed out on the white satin damask couch, naked from the waist down. Her sense of modesty is increasingly labile. I served her a late lunch, grateful to face only the blouse-covered upper half of her torso across the table. After I delivered the desired enemas, Ruth slept the rest of the afternoon.

Ruth's niece and nephew had invited us to dinner. I woke Ruth and helped her dress. She started with earrings and three necklaces, including crystal beads that once belonged to her mother and her favorite watch pendant, before choosing a two-piece blue and white print dress and white pumps. After trying on two others, she settled on her new short blond wig. I barely had time to wash my face and brush my teeth before Ruth was out the door.

After dinner, it was time for more enemas, then I helped Ruth into bed, gave her feet an oil massage, and brought her enzyme pills. The phone rang, and, while I was in the other room, Ruth vomited. It took an hour to clean her up, change the linens, and start two loads of soiled laundry. The incident made me wonder whether her Dilantin levels were low or whether the tumor was gaining ground.

Ruth's need for care is intensifying. I have to hold out until next week, when Olive and her friend Hyacinth, a registered-in-Jamaica nurse, will alternate ten-hour shifts during the week. The weekends are yet to be staffed. Ruth begged me to come back after Bob spends a week with her while I return to Colorado with Eben.

She pleaded: "I am afraid without you!"

"I know," I said. "I'm afraid, too."

<u>Day Twenty-Six</u>: Early this morning, I recalled a dream fragment about bears—the first dream memory since I met Ruth in New York. Even my unconscious has been displaced. By the time I leave Florida, I will have been gone a month. Bob and Eben will arrive late tomorrow night on my fifty-first birthday. Their presence is the only present I want.

The two males will stay at the Hilton by the beach, giving Eben the freedom he needs and Ruth the privacy she doesn't. There will be a week of extended family living as I orient Bob to his mother's care needs. When I leave to take Eben back to Colorado, either Olive or Hyacinth will be present twenty-four hours a day. In addition to providing Ruth with company and comfort, Bob will act as the Gonzalez protocol advisor. He won't be cooking, doing laundry, or helping with the enemas. I predict Bob will wonder why my long shift has been so hard for me.

Hearing is supposedly the last of our senses to disappear before death, but Ruth makes me think it is a close race between hearing and vanity. On the first night we were in New York for her appointment, as we got ready to go to sleep, Ruth leaned close to the harshly illuminated mirror above the hotel sink.

"I don't understand why I have all these lines above my lip. I never had those before!" she said.

I naively tried to sooth her, "Ruth, you are seventy-nine. They are normal signs of aging."

Since that moment, Ruth has puzzled over her lip lines at least five times a day, triggering a rare moment of exasperation in Olive.

"Ruthie!" Olive exclaimed, "God made you to get wrinkled so you would spend less time thinking about yourself and more time thinking about others!"

A much better response than mine, although it failed to detour Ruth from her fixation on creases. The other evening, low on disposable bed pads, we drove over to the new Eckerd Drugstore, where she added an expensive jar of face cream to our purchases.

One night while I was giving Ruth's feet an oil massage, I asked her an intimate question, "Ruth, do you and Maxwell have a sexual relationship?"

Smiling her great smile, Ruth said, "Well, he can't get an erection because he's an *alter kocker*[*], but he kisses me everywhere instead, which feels very good. In fact, I like it better in some ways!"

Although Ruth and Maxwell have been dating for almost a decade, both of Ruth's sons are in child-like denial about their mother's late-life sexuality.

Last night, Ruth was resting on her bed as I was putting away her clean laundry, when she suddenly pulled herself up into a sitting position.

---

[*] Yiddish for old shitter.

"I want to say one thing to you!" she spoke loudly, "I want to see you lay down and put your feet up whenever you can. I want you to stop all this writing!"

For a brief moment, before my brain absorbed the true text, I thought she had said, "I want you to stop all this laundry!"

Realizing her request, I said, "Ruth, I enjoy writing; it helps me to understand my experiences. I have been writing in my diary since I was a child."

Ruth's response was louder, more vehement, "I don't care! I don't like it! I want you to stop!"

Shifting down to neutral, I spoke more softly, "Ruth, I've given you all my energy and time these past weeks. My writing is the only thing I have from my own life."

Ruth's tone suddenly turned seductive, "Don't you want to please me?"

My spleen quivered. "I see what is happening! This is it! This is what Jewish mother guilt is all about! I am doing everything I can to please you. I am only writing when you are watching television or sleeping!" I was more than ready to fight.

Sensing this, Ruth quickly switched back to her *El Capitan* power stance, "I don't care about that! I must insist that you rest whenever you can. You aren't resting, when you are writing. Besides, I just don't like it!"

I stopped folding towels, moved to the bed, and sat down with my hip touching her thigh.

"I am sorry, Ruth," I said, "I am not going to stop writing, but I am willing to talk more about the feelings you have when I am writing."

Our standoff was interrupted by Selma's six o'clock call. While they were having their usual exchange, I wrote about ours. I actually doubt I will write much more while I am here. Bob and Eben will be welcome, time-consuming disruptions, but stopping is a risk; Ruth will assume I have agreed to her demand.

During my second week with Ruth, I found myself apologizing to Olive for both the content and tone of something Ruth had just said to her.

Olive responded with her usual wisdom, "Don't worry, Joanie. Ruth is used to being *El Capitan*, so it is hard for her now. Besides, I learned to live on the lower limb a long time ago. It don't bother me!"

For as long as I have known Ruth, she has been rude to any person who is paid to help her—waiters and waitresses, house cleaners, beauticians, security men, taxi drivers, and salespeople. Ruth commands what she wants, never expressing thanks. The disease process is exaggerating her habitual behaviors. Her manners, which were never great, were the first characteristic targeted by her tumor.

Living with Ruth has sometimes made me feel as if I was living with my mother again. It has been hard for me to keep an adequate distance from Ruth's powerful persona. Raised in the West, where social status has shallow roots in the wide-open spaces, I feel uncomfortable in the Florida caste system. Black and Hispanic women do most of the odious jobs, while white women sit around having their nails and hair done. Olive made a big fuss when Ruth told her about the weekend bowel accident. Both women responded as if I shouldn't have had to handle such tasks.

Ruth declared, "That's why I want Olive here all the time!"

I found that comment insulting to both Olive and me. Ignoring the idea that I might have been less competent than Olive, I persevered.

"Ruth, why should it be more appropriate for a non-family member to clean up after you than a family member?" I asked.

"You just shouldn't have to! That's all!" Ruth victoriously dismissed the topic.

I will leave Florida with extra baggage carrying the weight of this question. What is the *optimal distance* between illness and intimacy? Why do non-family members become preferred providers? When Jane was dying, I took care of her body instead of Tom, who had much greater familiarity with it. Who or what were we colluding to protect? The dying wife? The bereaved husband? Their romantic memories? Is Ruth protecting me or herself when she prefers Olive clean up her feces? I suspect she feels less dependent and more in control with paid help at her feet, than with her daughter-in-law, even if she is a *shiksa*.

Last night both Carrie and Harris called, each in pain and terrified by what was happening to them—stabbing abdominal pain, ascites, and night sweats. I have more intimacy with these symptoms than Dr. Gonzalez and have also felt the fear of death that accompanies such intense discomfort. We patients wonder if the symptoms mean our disease is advancing or are the discomforting sensations secret signals that Dr. Gonzalez's protocols are working? In the months before my remission, I was as desperately ill as I had ever been. Only the still vivid memories of my own agony allow me to speak to Harris, Carrie, and Ruth with any degree of equanimity. What I haven't admitted is just how afraid I am for all of us.

It was one o'clock in the morning in Hollywood, Florida when I said goodnight to Harris in Garberville, California. On the floor of the cat bathroom, I wept silently, too tired to treat myself, and bone lonely from keeping company with so much desperation.

-34-

Looking Back, While Flying Away

Day Thirty-Five: I am flying away from Ruth, leaving her in the care of Bob, Olive, and Hyacinth. Eben and I are on our way home to Colorado on a three-phase Delta flight, with plane changes in Orlando and Dallas/Fort Worth. Eben is in the window seat, eyes moving from cloud scenes to Legos. These distractions create time for me to write.

Just before Bob and Eben arrived in Florida, Bob sent a box of my favorite oriental lilies. Olive took the call from the security desk.

"A flower delivery is waiting downstairs," Olive reported, "It must be from your husband!"

After I insisted that Bob never does such things, Olive was delighted to be right. We made two birthday bouquets, one for Ruth's bedroom and one for the living room.

Seeing the bouquets, Ruth looked at me quizzically and asked, "Is it your birthday or something? I thought it was the twenty-first."

Ah ha! In her mistake, Ruth was astrologically assigning me to the cusp of Cancer. Later, I found Ruth rustling in the cupboard for a birthday card.

I begged, "Ruth, please don't!" and she reluctantly acquiesced.

The fact that Ruth had not prepared a birthday card for me in advance is a vital sign of her decline. Ruth and her friends have a great deal of self-esteem invested in card exchanges. There are hundreds of birthday cards bundled by year throughout her condominium. The cards are left out so that the Hallmark tributes can be read over and over again, like little acrylic love notes.

What happened in Ruth's generation to make Hallmark cards so intrinsically meaningful? The cards that Ruth received for her birthday this year are next to her purse, make-up, and jewelry on her built-in vanity—a spot as sacred as an altar in the bedroom of a Boulder Buddhist. Ruth repeatedly counts the cards. Based on her numerical narrative, this year she received in excess of fifty. But, no matter how many times she counts, two cards will always be missing—mine and Maxwell's.

On my birthday, Ruth got ready for Bob's arrival by insisting that Olive take her to Future Nails for a manicure, pedicure, and to get waxed. Ruth lost all her body hair during chemotherapy, except her chin hairs. Any woman who has been through chemotherapy knows that chin hairs are mysteriously impervious to the most toxic drugs.

Olive and Ruth were gone almost three hours. I slept through two of them on Ruth's white satin damask couch—a blissful birthday nap during which I dreamt of Ben Evans.

Ruth returned from Future Nails sans acrylic attachments—the first time in forty years she has been without artificial nails. I decided to take her abandonment of this particular artifice as a compliment, whether she meant it or not.

The ten days of my overlap with Bob were wildly challenging. Bob and I alternated beach outings with our son. In between, we found time to buy roller blades for Eben and underwear for Bob, who left his own at home. I judged Bob's packing omission an unconscious ambivalence about the task before him.

On his first night in Florida, Eben was adamant that I spend the night with him in the hotel because, as he said, "Mom, I need your full attention!"

I promised Ruth that I would return in time to confer with the Hospice nurse the next afternoon. However, after a morning on the beach with Eben, I struggled to restart my medical motor. Returning to Hillcrest, I growled at Bob's excuse for having forgotten Ruth's middle-of-the-night dose of pancreatic enzymes.

"I didn't have an alarm!" he pleaded.

A few minutes later, I snarled at Bob again because he had failed to read the instructions for the "Clean Sweep" protocol correctly.

"I didn't read far enough," he said, his defense skills growing steadily weaker.

My diagnosis: Bob's ambivalence about his nursing task was producing a precipitous drop in his cognitive functioning.

I gave the Hospice nurse a stone-cold look when she told me she had forgotten to obtain Ruth's Dilantin level from the lab. Then I completely fell apart when a second nurse in the Hospice office told me she couldn't get Ruth's lab results because she didn't know the phone number of the lab. Worse yet, she claimed to have no phone book.

I felt as if the whole cast of characters was not only incompetent, but conspiring to make me irreplaceable as Ruth's caregiver. I began sobbing. Seeing my exhaustion, Bob suggested we all go out for dinner.

When we returned, we found a frantic message from Carrie's husband on Ruth's answering machine, asking me to come to South Miami Hospital. Carrie had been taken by ambulance in a semiconscious state, extremely dehydrated with multiple blood clots.

I drove fast, trying to access my reserve tank of compassion and arrived just as Carrie was being settled into a bed in the ICU. Her husband, who had been drinking heavily, was being driven home by their son and daughter. Bob's ex-sister-in-law, Racine, who was not only Carrie's cousin, but also her best friend, along with Racine's husband, were with her. Racine introduced me to Dr. Cushman, the cardiologist who had agreed to follow Carrie locally after she chose Dr. Gonzalez's program.

I realized that Carrie had most likely gotten into trouble by continuing the enemas, even though she was too nauseated to eat or drink. I felt ruthlessly responsible, as if I had been practicing medicine without a license.

Feeding Carrie chlorinated chips of ice, I gently rubbed her back. When she tired of responding to questions from the nurse, I tried to explain something about the treatment Carrie had undertaken, feeling more foolish with each answer.

An hour later, Carrie's children returned to the hospital, numb from terror, carrying reversed responsibilities as guardians of both parents. They dutifully took Carrie's instructions for what they were to bring her in the morning before saying their goodnights, followed by Racine and her husband.

After watching the nurse give Carrie a shot of morphine for pain, I sat with her until she fell asleep, gently working pressure points on her back.

When Carrie had drifted away in the haze of morphine, I started down the hall to the elevator. Feeling as sleepy as the patient, I stopped at a vending machine for a caffeinated soda to keep me awake for my return trip. Approaching Ruth's car in the parking lot, I suddenly felt a strong urge to return to Carrie. I went back inside the hospital.

According to the clock above the elevator, it was twenty minutes after one o'clock in the morning. Carrie did not move when I entered her cubicle. I sat down next to her bed, watching her shallow breath, trying to understand what had brought me back to her bedside. Ten minutes later, there was a commotion as a team hooked up a badly injured car accident victim to the complex machinery of life support. Then a different nurse pulled back the curtain to Carrie's cubicle, syringe in hand.

"What are you giving her now?" I asked. Carrie already had two IVs, one for dehydration and one for heparin to dissolve the clots.

"It's a shot of morphine ordered by Dr. Cushman," the nurse responded.

I stood up quickly, "I don't think you should give her that! She was given a shot of morphine less than an hour ago."

The nurse looked surprised, "Are you certain? It hasn't been charted."

I pointed out the nurse who had administered the first shot and she moved across the unit toward her, syringe still in hand. After a quick consult and several furtive glances in my direction, she returned.

"You were right. I'm glad you told me," she said.

We both knew that the second shot could have slowed Carrie's respiration rate to a point of accidental euthanasia.

I drove back to Hollywood in a typical south Florida thunderstorm, weeping as hard as it was raining, overwhelmed by the close call. What force had turned me around? Would the outcome be better or worse for Carrie and her beloved children? Whatever time remained for Carrie would be filled with pain and suffering, yet I believed her children needed every possible moment with her. Was my decision to go back to Carrie's bedside synchronicity—one of those inexplicable fateful interventions by the universe? It is simpler, easier, and less anxiety-provoking to be a cynic. It is better to view it as a random coincidence rather than an unprovable reminder that we are linked to the fates of others, particularly when we are in hospitals. Once again I had been instinctually drawn to a mother in crisis.

It was three-thirty in the morning when I pulled into the parking lot at Hillcrest. The guard in the lobby was sleeping so securely it took thirty minutes of my banging on the glass doors and lobby windows to rouse him—a good thing I was well-practiced. I spent a few minutes snuggled against Bob and Eben on the sleeper sofa in the living room before I stumbled into the den and fell asleep on the trundle bed.

<u>Day Thirty-Six</u>: Eben and I finally arrived early yesterday morning at Denver's Stapleton Airport, hours late. Eben, who had been asked by his father eleven days earlier to remember where our car was parked, proudly led me right to it.

The dark outline of the Rocky Mountains grew larger as we approached home, the peaks spectacularly back lit by a summer sky filled with sparkling stars. Reaching the top of the hill that revealed the lights of Boulder in the valley below, we opened the car windows and turned up Annie Lennox's *Diva* album to full volume. The cool, dry mountain air rushed in as we sped down the hill, Eben and I singing "Little Baby Angel" along with Annie.

Marley and Magic greeted us with delirious delight. The koi began rising in our pond at the sound of our voices, the scents of the garden, the feel of familiar niches, and the devoted creatures circling around me felt like a desperately needed shower for my soul.

After I tucked Eben safely into his own bed, I listened to the messages on our answering machine. On June 27, 1993, while Eben and I were flying home, Harris Jerome Zidel died, nineteen months after being diagnosed with pancreatic cancer. Harris, age fifty-one, left behind his wife, Peggy Anderson, and his daughter, Sophie, born the same month and year as Eben.

The news of Harris's death made me so weak, I could hardly climb the stairs to our bedroom. Retreating to the womb of my bathtub, I sat soaking and sobbing until the water turned cold. Then I began shivering with grief and fear.

-35-

A Wedding and Whitehead Island

Day Eighty-Two: It has been six weeks since the death of Harris Zidel. I'm back in Florida, having arrived at Ruth's door a week ago. Today is Olivia's thirtieth birthday. All I managed to do was to send her a check in one of Ruth's abandoned Hallmark cards.

This is my third caregiving stint in Florida, but it feels like my thirtieth. I am here to help care for Ruth, but also to deprogram Olive, the last guardian of the Gonzalez protocols. Olive has found it hard to loosen her hold on hope.

After telephone consultations with Dr. Gonzalez, Bob and Joel concluded that their mother's decline was too steep to continue pushing her through the various regimens. With Dr. Gonzalez's blessing, the two brothers, joint holders of their mother's health proxy, have given up on the Gonzalez program—both struggling not to feel too guilty, foolish, or cruel.

During my short absence, the tumor in Ruth's brain ruthlessly expanded its territory. She is now suffering from total incontinence and is unable to walk. She rarely asks to get out of bed; her wigs gather dust; her jewelry and make-up, including her rouge, are untouched. Ruth's home is now dominated by medical equipment—a hospital bed, overhead hoist, tray tables, and a giant toilet chair. The seat of her wheel chair has been transformed into a shelf for a large bag of adult diapers.

Although Bob and Joel had told Olive of their decision to stop all of Dr. Gonzalez's treatments, when I arrived last week Olive was still force-feeding Ruth the Gonzalez diet. Ruth reacted joyfully when I offered her strawberry sherbet and apple sauce. I assured Ruth that she had eaten her last spoonful of fourteen grain cereal and drunk her last glass of carrot juice, truthfully explaining that Dr. Gonzalez had approved of these dietary changes.

In the passive ways that men often delegate difficult emotional work to women, it was my responsibility to help Ruth understand, that, even though she was still alive, she was not well enough to travel to Boston to see her granddaughter Gale married. She had been looking forward to Gale's wedding for many months. In June, Lauretta had taken Ruth to *schmatte* row to shop for a dress to wear to the ceremony. She had chosen a heavily beaded silk chiffon dress in her favorite lapis lazuli blue. The price was good, even for *schmatte* row. The dress would have been perfect, maybe even understated, for any South Florida wedding, but its elaborate beading made it geographically garish for an afternoon garden wedding in Boston. The same day Ruth bought the dress, Olive had hung it from the shirred curtain rod in the bedroom to inspire her patient to keep up her strength for Gale's gala wedding day.

It was not necessary for me to explain to Ruth that she was too ill to attend the wedding. She forgot whatever I said as soon as I finished speaking—the tumor has disabled all of her short term memory. Yet, Gale's wedding and the blue beaded dress were stored in the hard drive of Ruth's brain, not easily deleted.

Whenever Ruth was awake, only seconds separated her repetitive queries, "Isn't it time for me to get dressed for Gale's wedding?"

Ruth's repetitions produced a middle-of-the-night inspiration. The next morning, I told Olive we needed to dress Ruth as if she was going to the wedding. I wanted her to see herself wearing the magic dress, the matching satin pumps, and the wide-brimmed hat that Lauretta had selected to complete the outfit.

Dressing Ruth took our full-strength, plus an hour for Lauretta to put on the requisite make-up. But it was perfectly targeted therapy, even curative. Ruth looked at herself in the mirror and beamed. Then Ruth asked for Maxwell, who promptly arrived to admire his Angel. Ruth never mentioned Gale's wedding again.

A few days ago, after Ruth had another grand mal seizure, the Hospice nurse decided it was too risky for Ruth to continue wearing dentures. (Her two sons had no idea their mother had dentures until I reported the Hospice decision.)

This was the final denouement in Ruth's conscious relationship to Maxwell, as she did not want him to see her without her dentures. In a faint mumbled echo of her old *El Capitan* voice, Ruth announced that Maxwell was no longer allowed to visit.

When I explained to Maxwell why Ruth had asked us not to admit him for his daily visit, he was astonished.

"Tell her I will take out my dentures! I don't care about that! Doesn't she know how much I love her?" He was earnestly distressed.

I assured him that I would tell her, but I think even Maxwell knew that as long as Ruth was conscious, she would not relent. At first, I saw selfishness in Ruth's vain refusal. But later I remembered how I felt when I was bald and bloated during chemotherapy.

There had been many times when I had looked up at Bob, begging him, "Please, don't remember me this way!"

Ruth wanted to be remembered by Maxwell the way she looked wearing her dentures, dressed in her glittering silk chiffon dress.

Later I went up to Maxwell's condominium to try to comfort him. He welcomed me inside and made me a cup of hot water, apologizing because he was out of tea. I promised that while I was in Florida, I would arrange for him to come for a daily glimpse when Ruth was sleeping, but added a caveat. Because Olive and Hyacinth were both dedicated and role-oriented souls, I knew they would find it hard to break Ruth's rule after I left for Gale's wedding.

Before I went back to Ruth, Maxwell said, "I don't understand how you can take such good care of Ruth, when you have similar wounds. You are dying of the same disease. Who is taking care of you?"

I felt tenderly touched and I liked it.

Day Eighty-Three: Yesterday, after trying to comfort Maxwell, I drove to Coconut Grove to see Carrie, who had also been in a steady state of decline since she was discharged from the hospital in July. Carrie's fourteen-year-old daughter opened the door and escorted me up the grand staircase to her mother's bedside. When our eyes met, Carrie smiled broadly. Although she appeared to weigh less than ninety pounds, she had not shed her elegant manners.

"Oh, Joan, thank you so much for coming!" she said. "I am so happy you are finally here because I want you to meet my first husband, Pierre. He and Bob were such good friends!"

Carrie began motioning with her left arm for Pierre, who was invisible to me, to come around from the far side of the bed. After our introduction, Carrie continued to converse quite rationally with Pierre and me. She was having a vivid hallucination.

I had come to say goodbye to Carrie before leaving for Boston, but most importantly I had brought her some morphine suppositories. Earlier in the day when we had spoken on the telephone, Carrie had told me she was in unbearable pain, so nauseated she found it impossible to swallow the oral MS Contin tablets.

"Even if I manage to take one, I can't seem to keep it in my stomach; anything makes me vomit. I need your help!" she spoke in the language of desperation.

Carrie was suffering like many other vulnerable cancer patients do at the end of their lives. Dr. Cushman, her cardiologist, had little experience in end-of-life oncological pain management and morphine is still strangely taboo. Carrie's current husband had been adamant that he did not want the assistance of Hospice. Instead, he was belligerently over-dozing on alcohol to blunt his own emotional pain. Finally, Carrie's rapid decline had left her feeling hopeless, solidifying her reluctance to call Dr. Isaacs for advice, assistance, or a prescription for pain.

I asked Carrie if she would like me to help with the insertion of the suppository. Looking in the direction of Pierre, she asked him if he would mind stepping out of the room for just a minute. Then she instinctively turned on her side and pulled up her gown. After inserting the suppository, I turned to go into the bathroom to wash my hands.

Before I shut the bathroom door, I heard Carrie call out, "Pierre, please come back now and be with me again!"

By the time I finished washing up and began setting out the next suppository, Carrie was drifting into the deep sleep that comes with relief from relentless pain. Stepping into the hall, I called to Carrie's daughter, who promptly came out of her room. I showed her the suppositories, explaining that her mother would need help inserting another one in about six hours, as I entered the time on one of the dosing charts I had brought.

As I reiterated how important it was for her to insert the next suppository on time to prevent breakthrough pain, the sweet oval face of Carrie's daughter lost color and her eyes opened wide. Her expression was a mixture of adolescent angst and novice revulsion at the thought of tending to her mother in this intimate way—definitely not a fourteener's idea of adequate parental distance.

Hugging Carrie's daughter, I said, "I think your mother really needs a nurse. When will your father be home? Can you ask him to arrange for one?"

Her soft voice was strained, as she explained, "He has been drinking quite a lot. Do you think you could you call Dr. Cushman and ask him to arrange for a nurse? I don't know when my brother or my father will be home, but I think we need a nurse before it is time to give Mommy another suppository!"

Remembering how I felt trying to find a solution to my own mother's care at her age, I walked down the grand staircase to use the kitchen phone in private. Using my best unlicensed-to-practice tone, I called Dr. Cushman and asked him to set up a twenty-four-hour private duty nursing team, with the first shift to begin immediately.

I also asked that he arrange for a prescription for additional suppositories to be delivered to Carrie's pharmacy by messenger. Then I called the pharmacist, telling him to expect the messenger and arranging for payment and delivery. Finally, I called Racine, giving her my diagnosis of the situation, stifling my own sobs as I listened to hers.

Back upstairs, I asked Carrie's daughter if we could talk for a minute in her room. I spoke as gently as I could about what had been arranged. I also tried to explain what was likely to happen to her mother over the next several weeks, encouraging her to call me at any hour, if she or her brother had a question or concern. Even if I was no longer in Florida, I assured her that either Olive or Hyacinth would know how to reach me.

Back in Carrie's bedroom, I put the top half of my body close to hers, whispering my loving sorrowful goodbye into her ear.

Then I stood, looked toward the opposite side of Carrie's bed, and, following her gracious protocol, said, "Goodbye, Pierre. It was lovely to meet you. Please take good care of Carrie!"

Day Eighty-Four: Today I climbed onto Ruth's bed and lay down next to her, stroking the grey stubble on her head. The time had come for me to fly away to witness the marriage of her first-born grandchild. I began humming a lullaby that had once soothed Eben. After the lullaby, I struggled with a spotty, voice-cracking rendition of "Amazing Grace," the song I had resorted to a decade earlier as I stood above my mother's grave. Finally, I attempted a Hebrew chant used to call spirit escorts another friend had taught me. Ruth kept moving her head in an effort to see my face, but she has lost all but a tiny slice of her peripheral vision. My ragtag repertoire depleted, I fell silent.

A few moments later, in a raspy voice, Ruth spoke her last words, "Where are my boys?"

"They will come soon," I croaked.

This was not a total lie. Bob and Joel would travel back to Florida for her funeral, but not before. Each son had already bid their mother adieu, admitting that they did not want to see Ruth in a state of further deterioration. Their caring instincts had shifted from their mother to the preservation of their own maternal memories.

My legs seemed filled with lead as I walked away from Ruth, carrying my suitcase through the halls and lobby, out onto the broiling parking lot where a taxi waited. Olive, almost twice my size in both breadth and height, hovered above me, wiping her eyes.

As Ruth's last familial link, not only did I feel ashamed to be leaving, but I was also filled with fear. Part of the fear was about my own future. I boarded the plane to Boston feeling betrayed and abandoned for both of us.

When the plane landed, I called Olive from a pay phone inside Boston's Logan Terminal. While I was flying north, Ruth had fallen into a coma. The Hospice nurse had just left, having told Olive that Ruth would be comatose until her death.

Day One Hundred Eight: Gale's garden wedding was staged under glorious blue skies, close to the shade of Ruth's special dress. It was another marriage between two Gentiles. Gale married a soft-spoken, dark-eyed Catholic boy she met in Israel during her junior year abroad. Ruth would have played her role with pride, smothering any disappointment. Standing close after the ritual toasts, Bob and I were grateful Ruth had not died before the ceremony.

The next day, Bob, Eben, my father, and I drove north, crossing into Canada. Our foursome took two ferry boats across the Bay of Fundy to Whitehead Island, collapsing in a house offered by friends. Whitehead Island is a speck of land at the far edge of the Bay of Fundy where seventeen-foot tidal changes are the greatest anywhere in North America, producing a constantly shifting moon-driven distance between land and water.

Hard for tourists to access, Whitehead Island had neither hotels, nor restaurants with Early Bird specials. We acquired fish directly from fishermen, so delicious that no matter how much I cooked, there were never any leftovers. We read, napped, gazed silently out to sea, and, fueled by Eben's energy, picked wild raspberries, scouted whales, built sand castles, and walked among the gravestones of the island's ancestral residents.

Eben Valentine and his mother on Whitehead Island, August 31, 1993

Carrie died on September 1, 1993, our last day on Whitehead Island. Watched over by blue herons, we respectfully asked the forces of the universe to look after Carrie's son and daughter. That day, standing at the edge of the Atlantic, the deaths of Harris and Carrie converged, merged with Ruth's coma, dissolving all remnants of denial, the anchor on my grief. Not wanting to be the last one standing, I imagined myself floating away on the receding tide.

Eben took my hand and stood close to me as we boarded the ferry for our journey back to Colorado.

"It is going to be okay, Mom!" he spoke in a determined tone. "I promise you—everything is going to be okay!"

-36-

Ruth's Death and Burial

Day One Hundred Twenty-Seven: Eben had been back in school for two weeks before Olive called to say the end was near. By the time the three of us dressed and packed our bags, Ruth had died. Thunder storms delayed our connecting flight in Dallas, but Olive and her husband, along with Hyacinth and her daughter, showed amazing compassion. The Jamaican foursome waited with Ruth's body twelve long hours for us to arrive.

At the Levitt-Weinstein Mortuary, I dressed Ruth's body in the blue beaded dress she would have worn to Gale's wedding and her favorite wig. The mortician applied make-up, including rouge, to Lauretta's standard. Standing by the elaborate satin-lined casket, Maxwell affirmed his Angel was still beautiful.

On his tenth birthday, Eben stood before eighty of Ruth's relatives and friends. Speaking extemporaneously, Eben gave a loving tribute to his grandmother as if he were a Mormon child at a Testimony Meeting. Ruth would have been delirious with pride. I stood in for her.

Ruth's body was to be buried next to Eugene's in her Mother Sarah's Family Circle plot in Wellwood Cemetery near Farmingdale, New York. Under Judaic law this had to occur within forty-eight hours of her death. For this reason, after her funeral, we returned to her condominium to receive family and friends for only one afternoon and evening, instead of the typical *Shiva* (seven) days and nights. I thought of Ruth's broken three-section porcelain platter as I removed the plastic wrap from the hastily ordered deli food, delivered on disposable aluminum trays.

Late that night, after relatives and friends departed, the process of dismantling the remnants of Ruth's life began. Our self-imposed deadline was noon the next day when a plane carrying her body, her sons, her two daughters-in-laws, and her four grandchildren would depart Florida for New York City. Our efforts to disperse or dispose of Ruth's personal possessions were rushed because none of us felt capable of returning to Hillcrest immediately after her burial.

Sorting Ruth's lifetime collection of clothing, including over twenty-five handbags, we came across a mysterious outfit. Made of white nylon, it was a short and sheer ruffled baby-doll gown with nipple holes in the bodice and matching crotch-less panties. Her two sons were incredulous, but I loved the idea of Ruth dressing up in this outfit for Maxwell. The lengthy survival of Ruth's sexuality and Maxwell's impotent and tender adoration had come out of the closet.

Like the day of my mother's burial, there was also a steady downpour on Ruth's. I wanted fresh flowers to put on her grave, a self-serving Gentile gesture incongruent with Judaic tradition. After dropping Bob and Joel at the cemetery office to complete the paper work, I drove back to a nearby floral shop with Ellie, Joel's wife, where I inadvertently locked the only set of keys inside the rental car. Ellie and I had to walk several miles back to the grave site, arriving drenched by the rain.

As Ruth's coffin was lowered into the grave, the Jewish mourners threw clods of dirt down upon it, while I tossed in white lilies and roses.

-37-

The Alchemy of Gentiles

Day Three Hundred Seventy-Seven: Today would have been Ruth's eightieth birthday. While there are no Hallmark cards in the mail, Ruth's presence seems palpable.

Two months after Ruth's death, *Longevity Magazine* published an article by Peter Radetsky, a reputable science writer, who had interviewed me in October as one of Dr. Gonzalez's star survivors.* I had agreed to a telephone interview with Radetsky without considering the possible consequences. As soon as the November issue of *Longevity Magazine* hit the newsstands, our phone began ringing day and night, an onslaught of inquiries filled our mail box, and for many months afterwards strangers knocked on our door, arriving unannounced from near and far.

Some of those who came or wrote were desperate for any hope for themselves or for a loved one. Others were among those obsessed with a kind of fictional health. They wanted to know everything—from the brand of coffee I used for enemas to exactly what I thought about when I mediated, an oxymoron. Both the worried-well and dying patients raised issues of treatment perfection, asking whether the K-Y jelly used to ease the insertion of the enema hose might be poisonous.

Ruth should have survived; she would have come alive answering all such inquiries. But the onslaught drove me deep underground in my desperation to have more distance from cancer and death. Each inquiry and sad story had the preoccupying effect of keeping me from living. It was also hard because I remained an agnostic with respect to Dr. Gonzalez's treatment even though I was still receiving it.

---

* Radetsky, Peter. "The Doctor of Last Resort," *Longevity Magazine.* Nov. 1993, pp. 56-59, 106-110. At the time of publication it had been eighteen months since my radical remission.

Now, whenever I think of Ruth, I remember the small differences in our blood as we sat close in the dark at the edge of the Atlantic Ocean. My relationship with Ruth during the final year of her life helped me to more fully understand the deep need each of us has to be a member of a tribe or family. It took fearful desperation to herd the two of us together. Living with Ruth, I saw more clearly why catching a glimpse of self-resemblance in Narcissus's pool makes us feel less alone, further from death, more hopeful that some part of us will go on forever, in memory, if not matter.

I grew up deeply fearful of the human instinct that seeks special status by distinguishing differences, a need that frequently unleashes very destructive power, as manifested in yesterday's Shoah and today's African and Islamic genocides. Remembering Ruth, I think of what the ancient alchemists discovered—that only separated elements can unite. Separation must come first. Only then does the combination create something of greater strength.

– BOOK V –

ENDINGS

God, as in God the Father, is someone we human beings created in an effort to make ourselves feel special, loved, and under surveillance.

Frank V. Lieberman, 1956

-38-

Death Never Arrives on Time

Totems in home of author, Boulder, 2017

When death takes a child or young adult, those left behind feel death was particularly unfair in arriving so early. Those living in pain, as well as those caring for someone who is bedridden, unconscious, demented, and/or incontinent know that the wait for death can feel endless. When death is gluttonous, gorging its way through a disaster like a tsunami or a pandemic, death erases the meaning of individual lives and the faith of survivors. It is extremely rare for death to show up at the right time and in the right form.

In 1989, when first told death would come for me within six months, I was desperate to stay alive. I volunteered for treatments from both licensed and unlicensed healers. Something they did contained the cancer, but the costs were substantial. I was soon caught in a cultural and ethical void, a peculiar limbo—alive, but terminally ill, surrounded by the denial of death.

My first instinct was to embrace my five-year-old son. When he grew restless and squirmed free, I attempted to straighten up the loose ends of my life. My responsibilities included innumerable instrumental tasks—how was the world going to get along without me?

Six months until death was a very short time. I began bargaining to leave at the last possible moment by making primitive sacrifices in the Church of Chemotherapy with hourly offerings of various hair, skin, heart, and brain cells. I would have been a feeble prisoner of war, one eager to trade made-up secrets in exchange for survival. When my cellular sacrifices failed, my husband and I made a decision to stop treatment. I cancelled future commitments and sought substitutes to take on my most critical responsibilities. Only hospice-approved palliatives remained.

I prepared my last will and testament, assigned my husband all my power, and graciously received the gift of a burial plot. Finally, in what seemed like an endless list of last chores, I decided how to dispose of my remains by arranging for them to be transformed into cremains in a cardboard container, rather than in the more expensive cherry veneer model.

These tasks raised the question of what to pack for the trip. I assumed my essence would soon be smoke and ash, but what if some part of me were to arrive at another place? If so, what should I take with me? I held no membership in any religious group. The naïve faith of my childhood had been exorcized, my apostasy certified by excommunication from the Mormon Church. But the shadow of a large, full-bearded man, who kept track of my sins and mistakes with an elaborate, non-computerized system of black marks, still hung around in my unconscious—not to mention the possibility of running into my mother again.

In Logan, persistent Mormon missionaries had knocked on the door of the Dream House almost every week. My father politely turned them away, while my mother either ignored their knocks or slammed the door before they could utter a single proselytizing word.

In August 1956, a few miles after we crossed the Utah/Idaho border on our way to a new life in Montana, I asked my father a question about God. My mother, exhausted from the final cleaning of her Dream House, was sleeping in the back seat, so I rode next to my father, a place of potential intimacy.

"Daddy," I said, "I feel kind of sad to be leaving Logan, but I also feel relieved. Whenever I was at the Ward House, I always felt like I didn't really belong there—like something secret was wrong with me. I feel a lot of confusion about whether or not there is a Heavenly Father. Do you believe in God?"

After several miles of silence, my father finally responded.

"God, as in God the Father, is someone we human beings created in an effort to make ourselves feel special, loved, and under surveillance."

Then he signaled that our conversation about God was over by asking me to find a weather report on the radio.

In contrast to my father, my mother taught me that God definitely existed. But she warned, God had the slimmest possible margin on Satan, who might overtake Him at any moment if all precautions were not taken.

In 1961, Europe's grand cathedrals left me in awe. I remember standing in the Cologne Cathedral silently asking myself, "*If there is no God, what inspired humans to make such an edifice?*"

In 1962 at the time I decided to go to Africa, Mormon doctrine prevented all black men, as well as women of all shades, from holding the priesthood, which was the membership card one needed to receive guidance from God. It was the racial discrimination, rather than the gender bias, that caused me to seek excommunication from the Church of Jesus Christ of Latter Day Saints.

I was twenty-three years old and hadn't been inside a Mormon Ward for a decade. During my excommunication trial in Berkeley, the presiding bishop announced the consequence for my apostasy was that my soul was condemned to burn in hell for eternity, the ultimate maximum sentence!

In 1978 the LDS Council of Twelve announced that God had changed his mind. It was now permissible for black men to be admitted to the priesthood after all. I wondered whether this change meant a reduction in my own sentence was in order. Apparently a few fearful remnants of faith were still hiding out in my psyche.

In 1989, when I began to pack my bag for death, I was surprised to find the condemnation by those Mormon bishops took up a disproportionate amount of space. Should I bring a fire retardant suit or burn ointment?

Truthfully, I had nothing else to put in my bag for death except a dark cloak of instinctual fear. But as the weeks passed, I cautiously added a few miniature totems, as well as some loose semi-precious stones of memory, including a small velvet pouch filled with several ounces of basic hope in the sheer magic of the universe—particularly in its synchronicity.

By the time I finished packing, I had also tucked in my hand-written list of life lessons, which, among other things, reminded me with comforting certainty that my two children and husband would survive and rebuild their lives without me. So as the sixth month ended, I said my "goodbyes" in agony and moved toward the open door of my life.

I stood at the threshold with my soul raw and exposed, waiting for death, my small bag of fears, memories, and totem-like hopes at my feet. Death was such an important trip! My heart beat so fast I could hardly breathe, while everyone else seemed to be moving in slow motion, blind to the brilliant colors of the world, immune to its stunning sensations. I was in a state of awe, filled with gratitude for the myriad tastes and textures of the earth, for each human being, animal, plant, and inanimate object that I would never see again. Anticipated absence made my heart grow almost foolishly fond of life.

As I waited at the door for death, relatives, friends, and former clients came to say "goodbye." Others turned away, unable to face their own mortality or tolerate the awkwardness. Some left casserole offerings on the porch or wrote notes on blank cards because Hallmark has yet to make a product that expresses the complexity of this kind of permanent separation. My step-daughter Gale wrote me a letter thanking me for teaching her how to vacuum and draw her own bath. These written goodbyes became a poignant and hilarious collection documenting the odd assortment of indentations I had left on the world—providing me with both the pleasure and pain of previewing my own eulogies.

As the final minutes on the clock of waiting slowly turned into months, my adrenal glands stopped flooding my system with epinephrine and corticosteroids. My heart rate slowed, my legs began to tire, and my eyelids often closed without permission. I drifted into a series of involuntary naps in the ether of acceptance, the kind of sleep that overtakes you despite your best intentions. I would wake suddenly to find my mind stuck in the open wound of my mistakes. Pulling myself back into consciousness stung and filled my mouth with a bitter taste.

The oncologists and radiologists, while perplexed by death's delay, remained emphatic that nonetheless it was close. Since death was likely to arrive at any moment, it seemed prudent to remain near the door with my bag. But there were days when denial dominated and I returned to old preoccupations and tasks. Also Eben, like all young children, relentlessly demanded I carry on with my responsibilities as his mother.

We passed one developmental milestone after another. Eben's sixth, seventh, and eighth birthdays were celebrated with overly-elaborate parties, as if I was in competition with my successor. There were funerals before and after those birthday celebrations, as members of my breast cancer support group and others who had undertaken the same experimental treatments, fell by the wayside.

My suitcase remained on the threshold, but its presence became normative. Bob and I began to step around it with relative ease. My survival took on its own notoriety on Friday April 19, 1992, a day that was also both Passover and Good Friday. Scans of both my liver and brain were clear. I was in full remission. I put the suitcase, still fully packed, away in a closet.

Two years later, when the cancer cells began to regroup, as metastatic cells always do, I took the suitcase out of the closet and placed it by our front door. But, instead of waiting beside it, I kept going with my "new normal" routine and roles.

The dark cloak of instinctual fear, packed for the first planned departure, no longer seemed necessary. In its place, I packed a lacy slip of depression. I found it extremely hard to get back into treatment. So many of the friends I had made in the natural networks of cancer patients were gone. Trying to avoid anticipated pain of more loss, I resisted making contact with other patients. I had no explanation for why I was still alive. Worse, every day I was barely able to keep my head above water as I struggled to stay afloat in the treacherous sea of survivors' guilt. Like most isolated survivors, I struggled with the universal question: Why me?

Among other things, continuing to live was socially awkward. One day, while pushing a cart down a grocery aisle, the mouth of a Friends' School parent shocked to see me, fell open.

"My God! Are you still alive?" she exclaimed!

Part of the world had let me go and I had done the same.

The cost of treatment continued to pull my family further into debt. Not only was it extremely expensive to keep restocking my medicine cabinet, but metaphorically a great deal of energy was lost by leaving our front door unlocked and open to death year after year. Yet, I lacked the courage to refuse more treatment while there were medications capable of controlling the cancer, particularly as less toxic drugs came on the market, along with some that mitigated the most debilitating side-effects.

Everything I had first put into my suitcase was now worn and worthless. The totems and loose semi-precious stones, once infused with loving sentiments, had lost their meaning. The velvet pouch was empty of magic. The hand-written list of my life lessons was faded and torn, no longer legible. The world had lost its color and texture. I had to work hard to remember what I loved about life with such indiscriminate passion. This is how a long illness, coupled with unexpected survival, changes the experience of day-to-day life.

Whenever my feelings overflowed and dripped into my social milieu, everyone in the vicinity quickly moved to mop them up, acting as if nothing important had spilled, before deftly changing the topic. I cannot blame them for failing to see what had atrophied in my soul during the long wait, more than a quarter century of being half alive and half dead—one foot in the world, one foot gingerly feeling its way into the unknown. To hold the death of another requires one to face their own mortality—not easy, even for trained recipients.

For me, the proximity of death created a dialectic between numbness and hypersensitivity. During the first years of my illness, much of what happened seemed to have an element of synchronicity. But after a decade, I began to see my attempts to find meaning in the world around me, or evidence of something beyond it, as a latent phase of bargaining. With the rational atheism of my father, I often watched myself with derision.

Nonetheless, I could not stop looking for evidence of a greater force operating in our universe. Coincidences kept catching my attention. Something in me responded with joy to the strands of synchronicity—delighted and comforted by inexplicable connections between dreams, nature, and human behavior. These barely visible synchronistic strands were a psychological safety net that kept me from a free fall into meaningless nothingness.

Talking about—never mind writing about—synchronicity, is extremely difficult. I have a hard-earned doctorate in knowing how easy it is to sound crazy. Synchronistic patterns are subtle and constantly reveal new connections. The dreams of others are boring—only the dreamer comes alive in the retelling.

Once I knew a thousand stories of synchronicity. Now most have disappeared. That is what memory does: it erases the clues, just as it sometimes acts to illuminate them. Nonetheless, while my heart is still beating and my mind is still sound, I want to bear witness to my belief that our bodies are under the influence of powerful cellular phenomena capable of leading us to our deepest secrets.

-39-

The End of Remission

When someone is diagnosed with metastatic cancer, the question is not whether the disease will kill them, but when. Even after radical remissions like my own, the metastatic patient is not considered cured. Granted, sometimes there is another outcome. Death comes from treatment side effects or other co-morbidities. Bob and I knew my remission was only a temporary reprieve from death, but a disruptive plan for the property adjoining our Marine Street home led us to behave otherwise.

While I was caring for Ruth in Florida, Friends' School relocated the Preschool from the little house next door to a building adjoining the new Primary School six miles east. The consolidated campus reduced administrative costs and facilitated transportation for families with children in both age groups. As soon as Friends' Preschool vacated the little house, the owner obtained a building permit to convert it into a seven-bedroom duplex for university students, asphalting the backyard to provide the requisite off-street parking. Bob and I put our home of twenty-five years (which had undergone two extensive remodels in 1975 and 1982) on the market asking our listing realtor to help find us something smaller.

I had a dream that led us to a property five blocks west. It was located at the base of Flagstaff Mountain on a short one-way street marking the western boundary of Boulder. We bought the lot, signed a one-year lease on a rental home, and began building our own dream house as if we had all the time in the world. Indeed, like so many others of our species, we designed it as if none of us, not even Eben, would ever grow older.

Four months into the twelve-month construction process, one of Eben's teachers told me she had been diagnosed with Stage III melanoma. As I looked at her young face, her eyes filled with tears.

Choking back emotion, she said, "It started on the sole of my foot, but has now spread. I heard you have had some good results with alternative treatments. I was wondering, could you suggest someone?"

As she spoke, the image of a dark mole on the sole of Eben's foot filled my visual field. Three days later we learned Eben had Stage I melanoma. His teacher's inquiry turned out to be a life-saver.

Like Thijs DeHaas, Eben was referred to Dr. Robinson at University Hospital in Denver, who promptly called in a plastic surgeon. The surgeon explained he would need to remove a sizeable amount of tissue from the sole of Eben's foot to ensure a clean margin. A successful healing process required Eben not to put any weight on his foot for at least a month. Since Eben's classroom was on the third floor of a building without an elevator and he had yet to enter puberty when hormones accelerate all kinds of growth, we took a risk by postponing the surgery for two months until school ended while we carefully checked his mole for any changes.

My final denouement with Dr. Gonzalez came a few weeks later. The first sign of trouble had been the sudden bleeding in Yellowstone. An exam by a new Boulder oncologist, Dr. Helen Goldberg, revealed uterine tumors even I could see on the ultrasound. I had told Dr. Gonzalez of increased pain, bleeding, and bloating, but he had said my tests showed nothing of concern. At Bob's insistence, I returned to New York two months later with the results of my conventional tests.

A process server passed me on my way into Dr. Gonzalez's new suite of offices, triggering doubtfulness. Dr. Gonzalez never looked healthy to me, but on that day his face was unusually grim and grey, surrounded by the scent of stress. Dr. Gonzalez reviewed the ultrasound and blood reports, wondering aloud whether there had been some kind of error on my last hair analysis.

Then he said, "I am going to call Joan right now and ask her to recheck it."

Had Dr. Gonzalez spoken my name by mistake? Or was he was referring to the magician who was able to detect cancer levels in a strand of hair? Was her name was the same as mine? Among my many sources of skepticism about his therapy had been the mysterious practice of determining the state of various organ systems by analyzing a strand of a patient's hair. Dr. Gonzalez claimed it made no difference whether the hair was dyed or permed. Apparently such chemical processes did not distort test results, but if patients drank or bathed in chemically treated tap water, they might never recover.

Dr. Gonzalez dialed a number; I overheard a female voice answer.

He said, "I want her hair sample from March re-checked as soon as possible since the patient is in my office," before excusing himself.

Twenty minutes later as Dr. Gonzalez reentered his office, the phone rang. Was his hair tester also telepathic?

An atheistic voice deep within my psyche whispered, "*Maybe it's just a fake call from a front desk collaborator.*" Whatever faith had carried me to this point was rapidly dissipating.

"Yes, what is her level?" Dr. Gonzalez inquired into the receiver. "That's a couple of points lower than you reported in March." (Listening pause.) "I understand. Thank you for checking." He hung up the phone, and said, "Well, it turns out that there was a greater drop than I thought. I want to adjust your program. You haven't started eating any soy have you? It is loaded with estrogenic stimulants!"

While assuring Dr. Gonzalez no soy had passed my lips, I also reminded him that my tumors were both ER (estrogen receptor) and PR (progesterone receptor) negative.[45] I no longer remember what Dr. Gonzalez changed in my program. I hailed a cab to LaGuardia and never returned.*

In Boulder, Dr. Goldberg sent me to Dr. Molly Romary, the only OBGYN surgeon on the provider list of my new insurance company, Colorado Uninsurable Health Insurance Program (CUHIP).[46] After Dr. Romary finished a thorough physical exam, I looked into her beautiful azure eyes and asked the kind of question reserved for rude patients.

"Dr. Romary," I probed softly, "if you don't mind, could you tell me how many total hysterectomies you have performed?"

Responding without hesitation, Dr. Romary said, "Six."

"Great!" I said, "Seven is my lucky number!"

Dr. Romary operated two days later, adding my uterus, remaining ovary, fallopian tubes, and cervix to the body ballast I had already thrown overboard in my efforts to keep my life boat afloat.

Eben and I ended up in cancer recovery mode at the same time. He was amazingly agile on crutches and back on both feet by the Fourth of July. In

---

* Dr. Nicholas James Gonzalez died suddenly of unknown causes on July 21, 2015 at his New York home. He was sixty-seven at the time of his death. Since the death of Dr. Gonzalez, his former partner and ex-wife, Dr. Linda L. Isaacs, has continued to provide treatment to patients using individualized nutritional protocols.

contrast, his mother was sleepless, irritable, and aphasic. For many years afterwards, Bob joked that my total hysterectomy with oophorectomy was hardest on our building contractor, an extremely kind and gentle man, who must have come to dread our budget meetings. Menopause is hard; surgical menopause is brutal. For me, it was also a glimpse of what I subsequently learned my mother had experienced much earlier in her life, at the time of my birth.

Following the surgery, I had a "highly suspicious" mammogram. Dr. Goldberg advised bilateral mastectomies, but first she ordered a new genetic test for mutations associated with breast cancer developed by Myriad Genetics, a Utah company.[47] The test results made me eligible for a Stage I clinical trial for an experimental drug labeled "anti-HER2"—now known as Herceptin.

I was re-baptized into the Church of Chemotherapy, but it was a new reformed branch with doctrines based on targeted molecular therapy. Science was staying one step ahead of me.[48]

-40-

The Substitute Wife

On Mother's Day 1994, as I lay in my bed in our temporary rental home recovering from Dr. Romary's seventh complete hysterectomy, I accidentally blipped into an NBC television movie, *The Substitute Wife*. In this pioneer film, set in Nebraska in 1869, Amy Hightower (Lea Thompson) is living with her husband Martin (Peter Weller) and their four children on a remote ranch. Learning she is going to die from a large tumor in her abdomen, Amy knows her husband is going to need a substitute wife.[49]

Even though he doesn't want to face reality, Amy is a practical pioneer woman who wants to protect her four children. So Amy goes into town to recruit her own replacement at the local brothel. Boy, was she ever in luck! Among the women is a prostitute named Pearl, played by Farrah Fawcett. Amy persuades Pearl to give up a life of prostitution for a life of substitution.

Martin Hightower is initially reluctant to bed Pearl, but as the family moves into terminal-relay-style polygamy, not surprisingly, Martin slowly manages to find it tolerable. Of course, as soon as he does, Amy has a miraculous remission! (Somehow I missed the clinical trial inspired by this story.) One can only guess whether the source of the mysterious remission is the wife's competitive instincts or the helpful respite provided by an extra housekeeping hand. But when Amy's remission comes to an end and she dies, Pearl has already been fully integrated into the family and becomes the perfect substitute wife.

Personally, I thought it was more important that she be the perfect substitute mother. When I watched this movie, I was feeling extremely replaceable and anything but heroic. The best I had done was to come up with a short list of women whom I forbade Bob to choose as my replacement, having deemed their mothering skills closer to a D- than a C+.

Or as I had said to Bob on several occasions, "I want you to remarry as soon as possible, but please promise me under no circumstances will you ever marry Helen Doe, Nancy Unknown, or Molly Not-Okay," all pseudonyms for women in our social surround who were waiting in the wings, sending signals that they were more than willing to take over my role.

My worries about substitution were not so much about what my absence would mean for Bob, Olivia, and Eben, but more about what was happening to Jane's children and Tom. The summer after Jane's death, her first- born child, Christina, was married in a ceremony at a rented garden a few blocks from my mother's grave. She was attended by her five sisters. Adam was the ring bearer. The bride and groom honeymooned in Hawaii; both had lost a parent to cancer—the bride her mother, the groom his father.

I had rarely seen Jamie, Jane's second-born daughter, while Jane was alive, even less during the months before her death. But one afternoon Jamie made a brief appearance. I followed Jamie outside and told her that time was running out for her to say goodbye to her mother. As we stood on the porch, I took both of her ice-cold hands into mine and explained my fear that if she didn't try to make peace with her mother, she might regret not doing so for the rest of her life. Jamie said it was too hard because there was always someone around. I promised to create time alone for her. On the agreed upon day and the agreed upon hour, I sent Jane's mother Jeanne home early. As soon as Jamie arrived, I made my own exit. I never knew whether any reconciliation had occurred, but was relieved when Jamie chose to join the family the night Jane died, as well as at the funeral.

Sadly, soon after Jane's death, Jamie disappeared into the depths of disordered eating. Rarely seen by Tom or her siblings, she had begun living with her boyfriend the year before Jane died. After Jane's death, the children found only left-over evidence that Jamie had been at the house—empty cereal boxes, dishes sticky with cereal flakes flavored with Tabasco sauce, an empty ice Dryers ice cream carton, the smell of vomit in the bathroom. All were signs on the trail of one daughter's desperation.

Tara, Jane's third-born child, was most drawn into the role suction that follows the death of a mother. She was now the eldest of the children still living at home, so she most often stood in her mother's stead. As the designated driver of the younger children, Tara lent her ear to their needs and complaints and approached housework with the same casual élan as Jane. Miraculously, as the unacknowledged house mother, Tara finished high school while earning a small income making jewelry.

Lizzie, like Jamie, did not finish high school. Early in 1992 Lizzie had moved into the home of a close friend's mother. She moved back home just before Jane died because, with Christina and Jamie living elsewhere, she finally had a room of her own. Lizzie had a real need for order in her life.

After Jane's death Ronald found comforting companionship with his first girlfriend, Ivy. Wanting a new start as he began high school, he told Tom that he was going to move out of the house into a small apartment with Ivy. They shared the rent with a third friend. To support himself, Ronald worked full time at theaters and video stores until he graduated with his Boulder High School class in 1997. He was the star of the senior talent show, playing his own piano compositions by ear, never having had a formal lesson. Among his five hundred fellow graduates, Ronald must have been one of the very few to have reached maturity—having climbed a very steep motherless mountain.

Katherine became the symptom-bearer of the Help children's collective grief. As the first anniversary of Jane's death approached, her asthma worsened and threatened her life. Katherine was blue by the time the ambulance arrived and spent four weeks in the hospital. I spent so much time with her in the ICU that I was often mistaken by other patients as a hospital staff member. The following year, and for several years after, Katherine had to be hospitalized in the late fall as her birthday and the anniversary of her mother's death approached. In 1995, she had an asthma attack so severe that she was put on a ventilator for six weeks. When the vent was removed, Katherine was transferred to National Jewish Hospital for three weeks of rehabilitation and continued monitoring. When finally released, she came home to our house where I cared for her until spring.

Addie was well-supported by the close community at Friends' School, where there were classmates, teachers, and parents she had known for years. This helped Addie to keep moving forward. She developed an interest in horses and fortuitously found her way to the heart, home, and commercial stables of Joy Berren, where Addie's instinctive love and respect of animals was noticed. Nuzzled by horses, stimulated by learning how to care for them, and watched over and cared for by Joy, Addie found a second home. She spent many nights at Joy's and frequently traveled with Joy and her daughter to horse riding competitions.

Adam was also surrounded by thoughtful teachers and parents at Friends' School. When he was home, Adam's sisters were loving and attentive and he made his own nest in Ronald's old room. He showed few obvious signs of distress, but no one believed Adam had survived the loss of his mother without scars.

Duncan and Denevere had a very hard life after Jane died. The house was more often than not empty during the day. They were lonely and, as Addie's homework and interest in horses increased, the two Pulik were walked and played with less frequently. They often escaped as soon as the front door was opened and could be seen bolting down the street towards a dry creek bed like a matched pair of race horses. Watching them disappear, it was often difficult to tell whether they were one or two dogs. Tom found himself bailing them out at the Human Society and listening to angry complaints about their barking on the answering machine when he came home at night.

Tom did his best to cope, but he could not attend to all the feelings, basic needs, or time-consuming demands. It was even harder after Jane's mother Jeanne moved back to California. Tom slowly moved off casserole welfare and into fast food ordering, only occasionally cooking from scratch. He tried to organize the kids to help with the chores, but eventually gave up. Meals were irregular; the contents of the refrigerator unpredictable. Statistically it seemed highly unlikely that Tom would ever find a substitute wife, let alone a substitute mother for eight children.

I think the kids might have starved had it not been for Andra Beach, whose son, Evan, had been Adam's friend since they were two years old. In addition to providing invaluable support to Adam and the family, perhaps the most important thing Andra did was arrange for the young female cook at Friends' School to use her creative cuisine talents once a week in the Help's kitchen. She would spend the afternoon making several entrees for the week, while Addie, Katherine, and Adam absorbed her gentle mothering presence like three pieces of dry toast in a bowl of warm milk.

-41-

The New Normal

As we were building our dream house, Margaret Martin Clifford was dying. Margaret had fled Boulder to put more distance between herself and the pain of her husband's infidelity. Returning to her childhood home in Santa Cruz, she moved into the home of her mother and step-father. She also had the love and support of her sister Cheryl, who also had metastatic breast cancer. Her father and brother lived in the area as well.[50]

In July 1994, Margaret made a final trip back to Boulder with her brother to gather the household effects most important to her before retreating again to Santa Cruz. As Margaret explained to me over a long lunch, she dreamed of finding a little rose-covered cottage, a modest place for Katie and herself, but this had to wait until she and Glenn had reached a financial settlement. Margaret's son stayed behind with Glenn to finish high school. Separation from her son was a barely tolerable side-effect of Glenn's betrayal, but Margaret felt unable to survive without Katie's sweet balm.

Sadly, Margaret did not have time to find her rose-covered cottage. The metastatic cells in her lungs quickly multiplied; she died September 19, 1994, only six weeks after her second retreat to Santa Cruz. Twin genotypes, her sister Cheryl died two months later.

Like Margaret's and mine, Cheryl's thymus had also been treated with radiation. All three of us were considered by some insane medical fad in the 1950s to be growing too big and too fast for our future femininity as we entered puberty. The thymus treatment only worked for me. I was the tallest girl in my sixth grade class for six short months. Margaret and Cheryl were gracefully tall all their adult lives.

On Thanksgiving Day 1995, to mark the first anniversary of living in our dream house, Marley and I hiked up the Viewpoint Trail on Flagstaff Mountain. At an overlook where the roof of our house was visible below, I hung an angel ornament on a large elm tree struggling to survive at the side of the trail. Speaking into the thin cold air, I thanked all the invisible forces of the universe for bringing us safely through the year, asking them to look after the children of Nuhiela, Jane, and Margaret. Then irrationally adding the dying elm tree to my request.

Over the next weeks other hikers began hanging their own angel ornaments on the tree. Soon the "Angel Tree" became a holiday destination for families and visitors. News stories caused the local chapter of the American Civil Liberties Union (ACLU) to threaten legal action based on a claimed breach of the separation of church and state. The ACLU Board urged the City's Open Space Board of Trustees to remove the ornaments. The Trustees agreed; but when their decision resulted in numerous protests, they announced the ornaments would remain on the tree through holiday season.

Almost immediately, all the ornaments disappeared. According to the *Boulder Daily Camera*, a group calling itself the "Party of God" claimed responsibility for "de-angeling" the tree. Anonymous hikers quickly redecorated it with other angel ornaments and the conflict drew television news crews from around the country. Sometime after New Year's Day, the Open Space rangers removed the second set of ornaments.

American culture has always had irrational boundaries between symbolism and faith. Since 1947, the City of Boulder has allowed the Flagstaff Star to be lit on the side of the Mountain, not far above the Angel Tree. Maintained by the Boulder Chamber of Commerce it illuminates Boulder's western façade with holiday cheer. Throughout the brouhaha, only a few friends and neighbors knew of my role because I was ashamed by what my violation of the Constitution had spawned.

In the spring, in an act of private reparation, I planted an elm tree, an *Ulmus Camperdownii,* in front of our dream house. The tree is a rare cultivar which cannot reproduce from seed. It has a broad crown and a contorted, weeping habit—a shape that reminds me of a hovering angel. Every December our *Ulmus Camperdownii* becomes our own "Angel Tree" when we decorate it with angel, animal, and bird ornaments, an agnostic Gentile tradition on private property.

For me the most wonderful aspect of our new dream home is that the western boundary of our property merges with Boulder's permanently-protected open space on Flagstaff Mountain. We see almost as much wildlife as can be found in the Hayden Valley of Yellowstone, just not all at once. Bears, cougars, foxes, coyotes, raccoons, skunks, rabbits, and bobcats are frequent visitors. Once a

Canada lynx circled our house and peeked through the French door into our kitchen.

Our avian neighbors include owls, hawks, eagles, peregrine falcons, turkey vultures, crows, magpies, blue jays, woodpeckers, blue birds, meadow larks, chickadees, humming birds, finches, and my mother's favorite, the ubiquitous robin.

The Angel Tree in front of the dream house

-42-

A Lethal Human Pathogen

People often describe their experiences on September 11, 2001 by recalling what a beautiful fall day it was in New York City, how blue the sky was, how benign the world appeared before everything changed. I could say the same about the day I met Patty. It was a beautiful September day in 1987, the sky was blue, cloudless. I felt energized by the idea of returning to the role of student. I had no idea that I was about to be exposed to a dangerous weapon, one which would eventually make me into a vector for a lethal human pathogen that would infect Jane's children.

Disastrous events remind us of how often happenstance determines fate. Patty and I were strangers; our paths crossed that September day only because we ended up standing in the same line. Both of us were registering for classes at Naropa University. The President of Naropa University introduced us; I knew him socially and Patty was doing development work for him in exchange for tuition.

Patty was short, but not slender; her brown hair had been tinted red, cut short. Highly adept at making conversation, Patty invited me to have a cup of tea in the sunny Naropa courtyard. I was signing up for meditation and journal writing classes at Naropa University to shape up my mind which had regressed into the ABC's of late motherhood, while simultaneously trying to sail through the initial cancer scare. Patty was more serious, trying to complete a bachelor's degree.

We ended up in the same journal course where the writing assignments stripped away any anonymity carried into the classroom. Initially, I found Patty's narrative only interesting, not dangerous. I would have described Patty as charming, intelligent, and unusually ferocious about getting what she wanted. Yet, I felt a vague sense of discomfort whenever I was around her. My disquiet came from a faint roar in the background, one that hinted of a rocky stream at peak flow, full of fury about the myriad obstacles in its way. Patty evoked the same fierce determination that I had often observed in my mother.

Over the next nine years, Patty's chaotic behavior did more to disrupt my life than breast cancer and my limited life expectancy. Before moving to Colorado, there had been a divorce from her first husband, with whom she had three grown children. Patty had immigrated to Colorado in search of a new life. She initially obtained work in the development office of the Colorado School of Mines and joined the Science of the Mind Church in Denver. There she had met Peter, a divorced businessman, with two grown daughters. They were engaged at the time of our fateful introduction in 1987 and were to be married as soon as Patty graduated in May. During that school year we had lunch several times and she invited me to attend her wedding, which I missed because of illness.

After her marriage, Patty started calling frequently. Eben, who loved to answer the phone, began to announce her calls with annoyance.

"Mom! It's Patty! Again!" he would shout without covering the receiver.

Bob recognized Patty's self-absorption right away. He avoided most of her cross currents by leaving town. But the minute I picked up the telephone receiver, I was buried under an avalanche of complaints and suspicions. Patty sought neither advice, nor counsel. She simply needed to dump her disappointments, of which she had a seemingly endless supply. Patty's paranoia was equal to my mother's.

In January 1996, I was post-hysterectomy and back in chemotherapy, sitting by Katherine's bedside at National Jewish Hospital, when I made up my mind to delete Patty from my life. The prior year she had used funds from her second divorce settlement to travel to Australia, Bali, and Fiji. In Australia, she met a man on a train and immediately decided he was her soul mate. Two days later, she called me from Australia at three o'clock in the morning after he abandoned her at Ayers Rock. Listening to her I hoped that Interpol would be called to the scene of one of her rages and would quickly recognize that Patty needed to be committed or jailed.

While she was traveling, I began fantasizing about options for keeping her at a permanent distance: a mosquito with malaria, a strain of drug-resistant tuberculous, a bus accident, or a sudden tidal wave as she waded on a Bali beach. Perhaps she would die from a bite of a poisonous snake on Fiji (an option suggested by the ghost of my mother). Instead she returned to Colorado, got another job, lost it, and then joined an exploitive cult called Miracle of Love before becoming part of a pyramid sales program for a weight reduction product.

Desperate to put an impassable psychological desert between us, I telephoned Patty.

"Patty," I said, "There is something difficult for me to say and for you to hear. Every time you call me, the primary purpose seems to be to express your critical judgments of others. I have tried to listen with compassion, but my friendship with you has taken a toll on my health and on my family life. For many reasons I have decided to withdraw from any further contact. Please do not call or contact me again. Do you understand?"

My betrayal of Patty was as brutal as Nonny's was of me in 1967. I do not know how long Patty was able to tolerate my rejection of her—probably only a few hours, perhaps only minutes. My change of heart may have come as a greater shock than any of the many dealt her by the men with whom she had romantic relationships. Those rejections came after shorter intervals of engagement. Worse yet, in my pseudo-tolerance, I had listened to Patty grieve, cope, and move on when many of her friends and work colleagues, of both genders, had told her that it was over. In 1992, after a couple announced that Patty's anger precluded continuing their friendship, she called seeking absolution.

Sobbing, she told me something I already knew, "The pain and devastation of having a friend turn on you is indescribable!"

I had hidden my distress, my dislike, my dismay, and my disdain in order to avoid what? Something in my psyche must have known that it was too late that if I told the truth the costs would be very high, not only to myself, but to others. As usual, Patty responded by reaching out to others in search of balm for her badly bruised ego. Betrayal is not possible without trust and Patty had trusted me. Patty called the three men in my life whom she thought might be able to change my mind, make me withdraw my declaration, convert my rejection into a plea for forgiveness: my father, Roland Evans, and Tom Help.

As Bob was quick to point out, with a sense of vindication, Patty did not call him. Bob had been the first to recognize that Patty was developmentally disabled by uncontrolled narcissistic impulses. She was Trump-like, before Trump. My father was puzzled; Roland Evans was annoyed; and Tom Help was vulnerable. Seeing Tom become ensnared in Patty's web was proof that, not only is there no God, but that life on earth can be a kind of hell.

Soon after my call, Patty lost her job and retreated, moving to Buena Vista in rural Colorado in order to share a rental house with a cousin on the periphery of a summer camp for Boulder Buddhists.

A few weeks later, Tom stopped by to pick up Katherine, Addie, and Adam after a day of play. We chatted as he waited for them to gather their things and then, out of the blue, Tom mentioned Patty.

"I think Patty has the potential to be a stalker," Tom said.

I knew about Patty's calls to my father and Roland Evans, but it had never ever entered my mind that Patty would approach Tom. They had briefly met at our home on Christmas Day in 1993 when Patty attempted to sell Tom a weight loss product that was part of a pyramid sales scheme.

In a rare display of annoyance, Bob lost his temper in the middle of her spiel, saying, "Patty, not here, not now, not ever!"

But that day in our kitchen, Tom reported that Patty had been calling him frequently. I apologized, telling him to please not respond to any of Patty's telephoned inquiries. Nonetheless, less than a week later, Patty invited Tom to dinner because she was going to be in Denver over the weekend. Ignoring my warning, Tom accepted, sealing his own fate and the fate of his children.

Patty used what I had come to recognize as her standard seduction style—endless late night telephone calls and poetic love letters of the ilk I feel certain Tom never received or dreamed of receiving. After all, how likely was it that a fifty-year-old overweight widower with eight children and few assets was going to attract a woman who disguised herself as the petite and vivacious Patty?

It was Mother's Day, when I realized that Tom had swallowed Patty's bait and her hook was in deep. It was late afternoon when Tom pulled up in his jeep to deliver a handmade Mother's Day card from Katherine as I was watering our front garden. Standing by his jeep, he gave me an update on the children and provided the first clue that we were all doomed.

"By the way," Tom said, "do you mind telling me exactly what happened between you and Patty?"

My mind was racing as I tried to formulate a brief synopsis for him. *Over the eight years I have known Patty, she moved thirteen times, had eleven different jobs, one marriage, one divorce, and three affairs—really six if you count the number of times guys tried to run away, but she somehow reeled them back into her net. She joined a cult, tried a pyramid sales scheme, repeatedly asked to borrow money, and last, but not least, she undertook one serious suicide attempt. She is self-absorbed, destructive, paranoid, a magnet for sorrow, conflict, and chaos. I probably tolerated her for so long because my mother suffered from paranoid schizophrenia. I truly regret every single moment of time she took out of my life and my family's life.* But before speaking, I saw something in Tom's face and caught myself, saying only, "Why are you asking?"

"Well, Patty and I have sort of been seeing each other," he replied.

"Tom, I understand how charming Patty can be, but you must watch out for yourself. She moves awfully fast when it comes to romance. I have seen Patty go from zero to a hundred in less than twenty-four hours." My heart was racing. Making a desperate attempt to settle myself, I said, "Won't you please come inside for a drink so we can talk about what is happening?"

"I better not," he said, "I'm late getting back to the kids as it is."

I wanted to throw Tom to the ground, stand on him and yank Patty's hook right out of his mouth, while threatening that if he ever dared to have contact with that woman again, I would? Would what?

I tried another tactic, "Look, Tom, I understand that you are in a hurry, but I believe it is very important we find time to talk about this. I never imagined that you would be drawn into a relationship with Patty because of my decision to end my friendship with her. I can't tell you how sorry I am that my actions led to your involvement with her. But the one thing I learned about Patty is that she finds it extremely difficult to take 'no' for an answer. Patty will do almost anything to get what she wants or needs. Worse, she is not very conscious of what she leaves in her wake as her speedboat of need rides over any and all obstacles. Please be cautious—Patty is fully capable of destroying your life, not to mention those of your children."

On Memorial Day weekend, Tom drove to Buena Vista to visit Patty. Several weeks later, he took Katherine and Adam to Patty's place for an overnight visit. Addie missed the trip—traveling instead to Colorado Springs with Joy for a horseback riding competition. A few days after they returned, Tom came to the house to pick up his three youngest children who were outside playing with Eben. I used the small moment of privacy to try to reach Tom.

Katherine had already told me about the visit with her usual tactful reticence, saying, "She seemed like a nice lady."

Adam spoke with his stomach, "She made us a homemade pie!"

Addie said, "I'm glad I missed it."

Handing Tom a glass of juice, I said, "Katherine told me that you took her and Adam to visit Patty. How did that go?"

Tom's face turned pink, as he answered, "Cupid has struck and I am mortally wounded!"

"Oh, Tom, it is not my place to tell you what to do, but I think this is a terrible mistake. I am very afraid for both you and the children. Patty's romantic relationships always become filled with conflict and ruinous dissatisfaction." Tears were spilling down my cheeks.

Tom turned away, without answering, just as Adam came into the kitchen from the backyard. The conversation was over before it had even begun.

In desperation, I tried to contain my anxiety and guilt by telling myself that Patty would reveal her true persona before things went much further. Insomnia took over—my nights were filled with fear for the Help children, which made my mind dysfunctional during the day. Bob expressed his certainty that Tom loved his kids far too much to expose them to Patty.

Bob tried to assure me, "Tom won't put up with it. Don't worry! It will be over soon."

I wasn't certain I could put such stock in Tom. Patty's motivations were dangerously complex. She may have been in her standard infatuated thrall of early romance, but she was also on a quest to revenge the injury caused by my betrayal. So many men had been so innocently seduced by her before discovering that she was emotionally lethal. After the third year of knowing Patty, I had concluded her seduction tactics had to be ones that relied heavily, maybe entirely, on her providing male lovers with frequent fellatio.

In mid-August Tom stood in our kitchen, waiting for his three young children to once again gather their things together after a day of play.

"Patty and I are going to be married," he said. "We would both like you to be at the ceremony. It would only take fifteen minutes with Patty to put things right with her. We plan to marry in October."

My throat closing, I struggled to say, "Oh, Tom, I can hardly believe what you are telling me. You deserve to have the wonder of love in your life, but I have no faith Patty can provide what you and the children need and deserve. I wish I felt comfortable witnessing your marriage, but I can't. It would require too much social dishonesty. I hope you can understand and forgive me someday. Everything has happened so fast; my head is spinning. Have you told the children yet?"

"No, but I am planning on doing so today or tomorrow. Christina knows and the rest of them suspect something is happening."

"Look, Tom, I know I should respect your choice on this matter, but my nine years of experience with Patty are flashing blinding warning signals. I deeply regret that we never finished the discussion we began on Mother's Day. I want you to fully understand why I ended my relationship with Patty before you make such a permanent commitment. Could we have lunch tomorrow? Or I would meet with you whenever you can make the time. This is all so sudden."

I was talking to myself. Tom had already disengaged, his tone changed to cool and distant.

"I'll tell Patty what you have said," he said, as he turned and began gathering the kids' things.

A few days later, just as I was leaving for an oncology appointment, Tom unexpectedly knocked on our door. He had come to ask if I could help with a placement for Katherine at September School, a private alternative high school, and handed me her school records.

"Katherine is going to need a full scholarship, you know," he said. "By the way, Patty has moved in and Patty's granddaughter prefers Katherine to Patty and that is trouble." Tom spoke in a light, joking tone, as if the expressed preferences of Patty's granddaughter had no consequences.

I knew better and was frightened for all of them, but especially for Katherine. That was the last time I saw Tom before he married Patty.

I soon learned from Katherine and Addie that Patty's move into the Help family home had all the sensitivity and pace of an overnight corporate raid. Patty had the house re-carpeted and disposed of all the Help furniture on the main level, replacing it with her own. Duncan and Denevere were banished to the backyard, while the dog left over from Patty's second failed marriage, an irritable Shih Tzu called "Scout" was given free reign over the former territory of the Pulik.

In mid-September, Tara was sobbing when she called to report that she and Lizzie had been told to call before they went to the house. Patty had changed the locks on the doors to the only home they knew. Tara also reported that Patty was screening all calls; neither she nor Lizzie had been able to talk to the younger kids.

The wedding of Patty and Tom was held as planned at a Louisville restaurant. A week after the wedding, while Tom and Patty were away on their honeymoon,

I saw Addie and Adam at a friend's birthday party. They began talking simultaneously, hardly able to contain their tears and frustration.

Their tears were turning heads, so I tried to assuage their distress by asking, "Would you two like to go out for breakfast in the morning? Katherine is welcome to come too, if she wants."

Katherine, Addie, and Adam were waiting on the curb when Eben and I pulled up the next day. I waved to Lizzie, who had been allowed to come home for the honeymoon period in order to care for her three youngest siblings.

As soon as the five of us were seated in the Hotel Boulderado dining room, Adam asked me why I hadn't come to the wedding. I was gentle as I could be. As the three of them unloaded their rage about how they had been abandoned and betrayed, I found myself wanting to point out my own failures as a step-mother, wishing Gale was there to tell her own tales.

We shopped for new school shoes; afterwards we walked along Boulder Creek, using broken branches to steer golden leaves in the slow moving stream. Katherine, Addie and Adam elaborated on what Patty had taken from them: their older siblings, their dogs, their pictures of their mother, their sense of competence, and their self-worth. All their normal ways of living had been curtailed, ridiculed, or were now forbidden. Both their words and their body language told the woeful tale. Patty had taken them hostage in their own home and was stealing their life force.

We had a late lunch in the sun on the patio of a café on the Pearl Street Mall. Eben offered his sage advice to Katherine, Addie, and Adam, telling them he thought the three of them should run away.

Earnestly laying out his plan, Eben said, "All you have to do is to walk to the nearest pay phone and call us. There is one in front of Kmart. We will come pick you up right away and you can live with us! You need to bring Duncan and Denevere, too. Don't worry about your stuff! We can buy new stuff!" His advocacy was pure love.

The mystery in this terrible conflagration was Tom. Why wasn't he protecting them? Why was he allowing Patty to emotionally abuse his children? In Katherine's case, there was physical as well as emotional abuse. Both Tom and I had been through the education program at National Jewish Hospital. We had been taught to understand asthma's life-threatening risk to Katherine of environmental irritants like new carpets and dander-producing pets, not to mention the stress of so many changes.

The next morning, as I was showering after another restless night, someone began making hard hits with our front door knocker. Before I could put on my robe and wrap a towel around my wet hair, another round of hard knocking began. When I opened the door Tom stood before me—looking more polished and well-groomed than I had ever seen him But Tom also had a new, icy demeanor. Having returned from his honeymoon that morning, he had come to tell me that Patty wanted me to have no further contact with his children. I wept, pleaded, and begged him to not demand that I abandon them. He said that it was a choice I had made when I chose to reject Patty.

Lizzie later told me that she and the three younger children were waiting to greet Tom and Patty when they returned. Adam had started talking as soon as the newlyweds opened the front door.

"Joan took us out to eat and to buy shoes!" Adam proudly announced, as Katherine and Addie showed the honeymooners what was on their feet.

Saying nothing, Patty angrily stomped downstairs and locked herself in the master bath. When Tom followed, Patty screamed through the locked door that she would not come out until Tom promised her that I would never be allowed to have contact with the children again. Tom met his bride's demand.

Bob and Eben felt strongly that I had to let the children know what had happened, that leaving any explanation up to Tom or Patty was too risky. I struggled to compose a letter to the three children trying to explain why I would no longer be able to spend time with them and faxed a copy to Tom at his business.

A month later, two days before Thanksgiving 1996, Katherine was once again hospitalized and intubated following her third life-threatening asthma attack. The attack was probably inevitable. Any one of many changes could have caused it: the impact on Katherine's body of the stress produced by so much conflict, the anniversary of her mother's death, the fumes and fibers from the new carpet, the allergens spread through the house by the dander of Patty's dog, the loss of my support and companionship. The combination was a truly lethal threat.

When a friendly hospital nurse told me that Katherine was no longer on the ventilator, Eben called to ask if he could speak to Katherine. Patty came to the phone. Eben explained to Patty that he wanted to tell Katherine that he was thinking about her and hoping she was feeling better and he wanted to know what she wanted for Christmas.

"You may not speak with Katherine. We want nothing more to do with your family. Please do not call again!" Patty said before slamming down the receiver.

Patty was taking her revenge by punishing Eben. He wept for hours and, like me, he was grief stricken for months.

Six months after my banishment, Katherine asked Jeffy Griffin to take her to see me. When they arrived, I was away from the house walking Marley. I might have missed the moment if not for Marley, who incongruently stopped dead in her tracks and refused to go any further. The minute I turned around, Marley took the lead and pulled me towards home, where Katherine and Jeffy were waiting on the front steps. After that surprise visit, several times a year I would have the same reoccurring nightmare—dreaming Katherine was disappearing around the corner before I could see her.

Christina, who was now a mother herself, called me one day in 1999 to tell me Katherine was once again at the Mapleton Rehabilitation Center after another hospitalization for asthma. Christina clearly didn't know about my banishment when she suggested Katherine would love to see us. Eben was beside himself with joy to be able to see Katherine again. Addie and Katherine were siblings in his heart. Eben and I went to see Katherine at the Mapleton Center several days in a row before a nurse friend called to warn me that Patty and Tom had given instructions to staff that we were not allowed to visit. The nurse wanted to spare me the humiliation.

Once or twice a year, one friend or another would report on the latest Help news. That is how I learned that Tom and Patty had sold the house in Boulder and moved into a new home in Niwot, about fifteen miles northeast. Along with this piece of news came the knowledge that Adam was no longer attending Friends' School and that Katherine had been forced by Patty to withdraw from September School. Tara and Lizzie had been accused by Patty of vandalizing Patty's new garden, a reminder of Patty's peculiar paranoia. The last snippet of news that I heard about Tom and Patty included a report that they had sold Tom's business, as well as their new home, and had moved to Minnesota with Adam.

For almost a decade, my heart rate would accelerate whenever I caught even a glimpse of long blond hair that might belong to Addie or Katherine. I laughed at my frozen memory when a friend told me that Addie's almost white blond hair had been cut short and dyed black. Further, her pale slim body had been tattooed and pierced numerous times.

I was delighted when I learned that Lizzie had a daughter named Maddie who was a student at Friends' School. I felt close enough to reach out and touch a Help child or grandchild, but, as another friend sagely reminded me, the Help children were old enough to contact me on their own, if they wanted. Only my friend's repeated reminders restrained my impulses.

Just after Christmas 2004, I learned that Tom and Patty had filed for bankruptcy and the home they had bought in Minnesota was in foreclosure. Perhaps predictably, the house was located at the end of Hell Road. My friend also had what seemed like good news—Katherine was engaged to be married and had recently moved back to Denver from New York City. Then in February 2005, I received a special Valentine, a letter from Katherine.

February 10, 2005

Dear Joan,

I hope this letter find you in good health. I have wanted to make some sort of contact with you for a very long time. I have a stack of half written letters to you, but I always felt awkward because so much time has passed, and about the way my father and Patty treated you. Today I've decided I couldn't let any more time pass without letting you know how much you are appreciated and loved, and to thank you from the bottom of my heart.

As I look back on the time we've shared I am truly humbled by every thought I have of you. Because you gave your time, energy, advice, love, and compassion so freely to us. Every experience I had with you was magical. Some of my most cherished memories are of you. You gave me a childhood filled with laughter, adventure, and happiness that would have not been possible without you. I am so grateful to you for your kindness, compassion, love and strength and I need you to know that you are loved and appreciated. You hold a very special and cherished part of my heart

I also wanted to thank you for being such a great friend to my mother. I can't imagine what it would feel like if I were her to have had such a wonderful friend. I am sorry it has taken me so long to tell you how much you mean to me.

Love Always, Katherine

Katherine and I met at Tattered Cover Bookstore in Denver before sharing a meal at a nearby restaurant. At the end of our long afternoon visit, I wanted to insist that Katherine come home with me. But she was now twenty-four years old and living with her fiancé, Roger. He had grown up in Longmont and Katherine had known him since she was sixteen. She told me that both she and Addie had been asked to leave the new Help family home in Niwot before reaching their sixteenth birthdays. Needing to find self-supporting work, neither had been able to finish high school. Katherine had obtained her GED while working as a hotel clerk.

Two additional reunions followed my initial meeting with Katherine. At Easter, Katherine, Addie, and Tara joined us for dinner. On Mother's Day, Lizzie and Maddie were there too. Looking at the five of them seated at our table, I thought of how proud Jane would be of her progeny.

A month later, Katherine moved into our home, having decided that she wanted time apart from Roger to see their relationship from a distance. Bob and I encouraged her to try a college course through CU's Continuing Education Department. She agreed and I helped her enroll in a five unit summer psychology class. Katherine earned an A grade and was ready for more. We quickly rented a studio apartment for Katherine off campus and spent a frantic week getting the studio furnished and operational before the beginning of the 2005 fall semester.

Katherine took a full load of continuing education classes. Based on her performance, CU admitted her as a regular student and gave her a tuition scholarship. Showing remarkable intellect and resilience, Katherine finished her degree in three years, graduating with high honors in psychology. After Katherine presented her senior honors thesis, she was offered a place in CU's highly competitive neuropsychology doctoral program, a recognition of both her accomplishments and her future potential.

Tom unexpectedly drove to Boulder for Katherine's graduation. His progeny chose not to ask their father to join us for the family dinner at our home before the evening departmental ceremony. Because it was a school night, Bob took Eben home as soon as Katherine crossed the stage of Macky Auditorium, while I stayed until the end the ceremony. When all the diplomas had been distributed, Tom approached me outside the auditorium as I waited for Katherine. In a voice so soft I could barely hear him, he thanked me for helping Katherine obtain her degree. Nothing he said resembled forgiveness, but I knew

Tom well enough to know any gesture of gratitude was extremely difficult for him.

Katherine and I accepted Tom's offer of a ride home in his rental car. Both of us sat in the back seat, politely keeping our distance from a man neither of us understood.

-43-

The Most Eccentric Person in My Life

By
Eben Valentine Pelcyger
Casey Middle School, Boulder, Colorado, April 1997

My eighty-six year old grandpa, Francis Valentine Lieberman, is by far the most eccentric person in my life. He lives alone and does all his own cleaning, laundry, cooking, and shopping. He is a child of the Great Depression and does not have a credit card. He pays cash for everything.

My grandpa has worked very hard to stay alive. He cuts the fat off of every piece of meat before he cooks it. He weighs less than my mother (which really annoys her) and he still looks the same as he did when I was born fourteen years ago. As part of his strict regime, he takes five mile hikes in the mountains on Mondays and Fridays. Recently he joined a health club, so now he picks me up after school and we work out together. He is definitely the oldest guy at the club and the only one who doesn't own a pair of shorts or sneakers.

About three years ago, he put his television in his storage locker because he said there was nothing worth watching.

He says: "Since I got rid of the TV, I urinate less often!" because he is of the age when you have to think more than you want to about urinating.

For entertainment, he reads books and the New York Times "Large Print" edition. Sometimes he goes to the Library Auditorium to see old movies. We saw To Kill a Mockingbird because I was reading the book. He bowls with the Senior League and is happy when he bowls the highest score, sad when he doesn't.

My grandpa is a glue expert and has fixed everything I have broken in my life. I can call him anytime for a ride or for help with my homework. He still knows all the elements in the Periodic Table, even though he hasn't been a scientist since he retired in 1967. He makes his own bread twice a week. It is really good. Sometimes when I visit him, I end up eating a whole loaf. My grandpa has taught me some really important things about life and living. I am very glad that he is my eccentric.

Frank V. Lieberman on his eightieth birthday, August 26, 1991

In February 1997, my father was ticketed by an open space ranger for walking Marley off leash—an act which synchronistically saved his life for another fourteen months. Upset by the rule and its unfair application to Marley, who was always perfectly behaved in his presence, my father decided to write a letter to the Boulder County Commissioners, arguing that dogs under voice control should be allowed off leash on open space.

The tremor of age had made Frank's handwriting illegible, so he had stopped writing even his dutiful Christmas communications. But his ticket-triggered outrage led him to try Eben's retired Gateway computer covered with dust in his guest bedroom. He called me three times for help as he maneuvered his way through new technology. After the third call, I drove to his condominium to help him put the final touches on the letter, then took a copy back to our house on a disk to print.

In the morning, Bob asked me to drive him to work because he was late for an early conference call. Rushing to meet Bob's request, I left the house wearing my robe and slippers even though it was snowing, having picked up the printed copy of Frank's letter from the front hall table since Bob's office was a short block from his condominium.

Making my way gingerly across the snow-covered parking lot of the Friends' Condominiums in my modest robe and toeless slippers, I grumbled to myself, "*This is silly! Why don't I go home and Frank can get the letter when he comes to walk Marley?*" Although I had a key, for years I had been trying to teach my father to knock before entering our house. Modeling the desired behavior, I rang his bell. There was no answer. I could see his car in the garage. I waited a few minutes and tried the buzzer again. When there was still no answer, I broke my own rule and used my key.

I found my father unconscious on his bathroom floor, dressed in his coat, hat, and gloves for the wintery walk. As he was about to leave, he had felt dizzy and nauseated at his front door. He somehow managed to make it to his bathroom before passing out. I dialed 911. When the paramedics arrived, they could find no pulse. Unbelievably, forty-eight hours later Frank was released from Boulder Community Hospital with a pacemaker and yet another extension on his life lease. Both of us were gold medalists in the Close-Call Olympics.

Post-pacemaker, Frank was less energetic. His feet seemed heavier, harder to lift; his legs ached when he walked. Sometimes he relented when I asked him not to tackle a chore around our house. He told several of my friends I had fired him, complaining he had nothing to do.

"The Big Firing" took place in my dining room on the morning of my father's eighty-fifth birthday. The night before we had celebrated with a small family party featuring plain cake and reluctantly-received gifts. He had been working close to full time at our house since we moved in, tackling myriad tasks on an endless punch list and helping me with the landscaping. Our summer had been dominated by hard physical labor and the presence of a houseful of perpetually-hungry-fourteen-year-old boys. Like many end-of-summer mothers, I could hardly wait for school to start. Having dropped Eben at Casey Middle School, I came home to find my father already at work. Desperate for a pause from the Y chromosome, I tried to suppress my irritation.

"I really need to be alone today. I would like you to stop working and go home!" my voice sounded older than his.

"Why? I won't bother you!" he said. "What difference does it make if I keep working in the garden?"

As always, he was narrowly task focused. My father was thirty years my senior; I felt compelled to do the physically challenging work alongside him. Plus, his aging prostrate required frequent bathroom breaks.

Taking both of his hands in mine, I said, "It makes a difference. I can't explain why, but it does. I haven't had a moment alone for months and I am desperate for solitude. Also, I want you to devote your energy to your own interests, instead of ours. We deeply appreciate all you have done for us, but Bob and I need to be able to manage the house and garden by ourselves."

The result was a giant clarification conference. We were Yin and Yang. I felt I would die if I didn't stop working; he felt he would die if he did. We were trying to hang onto life using opposing strategies.

After that conflicted conversation, my father applied for his own library card and read through the Boulder Public Library collection of large type books. He also continued to perfect his bowling, hiked more, and spent less time helping our household. It felt a little strange to both of us at first, but we managed to make the appropriate adjustments, and, at least for me, developed a more *optimal distance.*

Four months later, Frank complained of angina. Dr. Turvey had retired, so I accompanied my father to several appointments with his new cardiologist. Reluctant to try any of the recommended medications, he finally began taking nitroglycerin before walking or bowling. By the end of March, however, he was having increased difficulty making himself comfortable, even with the extra nitroglycerin. He often had angina from simple acts like brushing his teeth.

His new cardiologist asked him to have another angiography. We called for an appointment, only to learn there would be a two-week wait. When my father finally had the angiography, it was too late for any balloons or shunts. Three cardiac arteries were completely blocked and the fourth was ninety percent occluded. The next ten days were hard. My father had reached the limit of interventional cardiovascular medicine.

Bob and I asked him to move into our home so we could provide him with more support, but he refused. Then the new cardiologist raised Frank's spirits by telling him that on a scale of ten, he was only on level two in terms of what medications could do to relieve his angina. His doctor promised to make him comfortable enough to resume his normal activities. It took another week of trial and error until the right dosage of channel blockers and nitroglycerin gave my father sustained relief for what turned out to be the last week of his life.

His mood was good because he was active again. On Sunday, my father joined us for a family dinner. On Monday, he went for a longer than usual walk with his old friend, Lenore Stewart. On Tuesday, he went practice bowling alone in Longmont. On Wednesday, he told me he was going "to rest" in preparation for his regular senior bowling on Thursday, except for his weekly grocery shopping (three stops for his favorite brands and prices).

When I spoke with my father on Wednesday night, he sounded more ebullient than he had in several years. He said he was feeling good and was looking forward to the next day for three reasons. First, his senior bowling team was in second place with a chance to overcome the first place team. Second, his grandson Eben would be playing his second-to-last league basketball game (my father was a dedicated fan, never having missed a game). And third, his beloved granddaughter Olivia was coming from New York City for a long weekend visit.

The next day the phone rang just as I was leaving to pick up Olivia at the airport. It was the chaplain from Avista Hospital in Louisville—my father had just arrived by ambulance in critical condition. Bob was away in Washington. Since cell phones were not yet ubiquitous, I had to take time to call two friends—one to pick up Eben and the other to page Olivia at the airport with instructions to take a cab directly to Avista Hospital. Leaving an anxious Marley behind, I drove fast on Highway 36, my heart pounding.

Reaching the Avista Hospital Emergency Room, I found my father was already on a respirator; his blood pressure was sixty over forty. He had been bowling with his team, The Hopefuls, when he felt pain. After taking more nitroglycerin, he became dizzy and vomited. He was trying to clean up after himself when he passed out. Coming back to consciousness, he told a female teammate to call neither me nor an ambulance, assuring her he would be fine in a few minutes. His teammate called the ambulance, but my father refused to give her my number.

"She's on her way to the airport. I'll call her later," he reportedly said.

At the hospital, Frank told the nurse, "I know what's wrong. What I need is an artificial heart!"

My father was joking and cooperative, regretting that his unfinished good game score would not help his Hopefuls team beat the first place Four Aces team. He was being wheeled down the hall for an x-ray when he complained of more pain. After his blood pressure dropped dramatically, he gave the ER doctor permission to put him on a respirator. I doubt my father believed his heart was finally giving up its long struggle to beat. Maybe he even imagined an artificial heart was a possibility. He was a deserving candidate in more ways than one.

The respirator made it hard for us to say goodbye. It required morphine for the pain, Midazolam to ease the respirator placement, and Dopamine to raise his blood pressure. The combination had put my father in a deep, drugged sleep. His limbs were already cold. Over the next four hours his blood pressure slowly dropped until it could no longer be detected by the machines surrounding him. He was in cardiogenic shock.

Looking at my father's face, I felt as if I was falling into a dark chasm, into oblivion. Having recently struggled to find greater physical distance from him, at that moment I was utterly incapable of imagining the loss of his perpetual presence.

Eben arrived by taxi, dressed in his basketball uniform and stood by his beloved BaBa, holding his hand until almost game time.

Then he said, "Mom, I know BaBa would want me to play. I think I should go and come back after the game."

Eben was right about what his BaBa would want. I gave permission with unresolvable reluctance and Avista's exquisitely kind chaplain drove Eben back to Boulder for his game.

Minutes after Eben departed, Olivia arrived from New York City. Six months pregnant with her first child, she was exhausted from a long day of travel. Only three weeks had passed since Olivia had been awakened by a dream about her Grandpa Frank. After her dream, she decided that instead of going to the annual meeting of the American Psychological Association in Boston, she wanted to come home—a dream-delivered decision for which I will always be grateful.

Olivia and I spent the next two hours talking to my father, holding his hands, rubbing his feet, and stroking his still thick, not completely grey hair. My father was unresponsive to our words, but his hands warmed in ours.

Game over and won, another friend drove Eben back to Avista. He arrived only moments after the nurse had removed the respirator—Eben's BaBa was dead.

Escorted by the nurse into the curtained cubicle, Eben said, "BaBa looks just like he does when he is taking a nap on his couch."

As we surrounded his body, I cut a lock of his hair. The attachment to the body is so strong. I needed a longer wake, but I had two very tired children and could hear my father telling me to "*go home.*" An hour later, I reluctantly drove away from his body with his two grandchildren and his first great granddaughter, yet to be born.

Bob returned the next morning on the first flight from Washington. Olivia and I met him at the shuttle stop and the three of us drove to the mortuary. There we awkwardly made the incongruent, impossible choices (e.g., Should my father's body enter the crematorium in an open coffin made of cherry veneer or in a cardboard box? Should his recovered ashes be placed in Carrera Classic Cultured Marble mini-coffin ($250) or an opaque glass jar ($25)?). Then we viewed my father's body one last time. The Crist Mortuary staff kindly admitted Marley to the viewing room. Bob lifted Marley up to the edge of the gurney so that Marley could smell for herself that her always-reliable hiking partner would never again come to take her into the sacred silent spaces of the Rocky Mountains.

My father's memorial service was held in the Friends' Condominium library adjoining his unit. Afterwards we served food and drinks to the mourners in his sixty-third residence. In addition to most of our blended family, his sister Alice Marshall, his niece Lynn and her husband, Robert Boehmer, a rich assortment of our friends, his neighbors and bowling buddies, and others like Thijs DeHaas, who knew and admired Frank V. Lieberman, joined us in honoring his life.

Eben had been stoic and controlled after his BaBa's death, but he had suffered a tremendous loss. Because Bob's work required him to be absent from home so frequently, Frank had truly been a *grand*father to Eben, always available to help with anything, showing Eben how to live a good and loving life. Eben's body soon spoke his sorrow. The day after the service, Eben became desperately ill; his temperature climbed so high so quickly, he had a seizure. Taken by ambulance to the hospital, he remained overnight and battled fever for seven more days until his immune system fought back and the mysterious virus finally moved on. In between, Eben collapsed into the kind of cuddling and tender care he needed, watching a video compilation of old family home movies over and over that had been sent by my cousin Julie Ann starring BaBa as a young man.

Several months later I began preparing my father's condominium for a new owner. He had lived in the Friends' Condominiums for twelve years—longer than any other place in his life. Shampooing the sea green carpeting, I wept as I slowly erased the map of his living etched in the wool fibers, the footpaths from kitchen to reading chair, to the north window overlooking the cottonwood trees along Pearl Street, just beyond the parking lot he involuntarily patrolled. On my knees, I used his special scrub brush on the tiny tea, tomato, and tuna stains barely visible between the leg marks of his solo dining chair. He was with me as I worked, providing the same critical coaching I had known all my life. My father's remembered voice was as powerful as ever.

Be careful now! That's too much cleaning solution for that size spot! Don't rub too hard! Just let it be, Mama, * *and come back to it. You need the upholstery tool for those edges. You better put some Liquid Gold on those window sills. Pour it on the cloth first—you don't want to spill it on the carpet!*

Compared to most, my father left behind a neatly trimmed field of possessions. Having moved so many times, he had already disposed of what most Americans are acculturated to collect. His life was dominated, indeed delineated, by moving. He remembered forty-seven different childhood addresses. This migrant pattern, initially driven by abandonment and poverty, continued as the USDA Bureau of Entomology moved him from state to state, from outbreak to crop failure. Then moving became his only remedy for his wife's mental illness.

Packing up my father's possessions, I was struck at how death reveals the debris of our good intentions. There are the buttons—saved, but un-sewn. The

---

* After Eben's birth, my father began calling me "Mama" to facilitate the development of Eben's language skills. Olivia always called me Joan because no one ever addressed me otherwise in her presence.

cards received—but not yet answered. The broken appliances—unfixable, and, worse yet for my father, not yet recyclable. (He left behind almost a dozen Sunbeam electric razors.) Then there are the patterns and practices of personal inventory—surpluses of StimUdent toothpicks, Pritikin canned soups, Northern toilet paper, Palmolive hand soap, Jergen's unscented hand lotion, and a tiny box of three mysteriously hopeful Trojan condoms.

Straightening up his kitchen in preparation for the memorial gathering, Olivia and I discovered three temptations in a kitchen drawer—three See's dark chocolate bars Eben sold to his BaBa eight months earlier for his outdoor education club. Two were still unopened, the third had been consumed except for one small piece. My father's self-discipline was extraordinary. I couldn't have been trusted to be alone with those for a week.

For thirty years, I watched my father struggle to preserve his life by steadily hiking through a deep canyon of caloric deprivation. His cholesterol was measured for the first time in 1968; it was three hundred twenty five, a number that marked the end of his relationship to American cuisine. As I worked to clean up his life, I kept his final chocolate vignette intact as long as I could, opening his kitchen drawer, imagining my father doing the same while holding his mug of breakfast tea, indulging in just one bite. His mother always liked a bite of chocolate after breakfast tea and so do I.

For years I had yearned for my father to be happy, to be in love with life—if not with another human being. But my father never learned how to make himself happy, only how to survive by avoiding conflict. A Gentile by both birth and cultural conditioning, his childhood experiences in Utah taught my father to keep a low profile, camouflaging himself with friendly helpfulness. A specialist in social discreteness, there were no bumper stickers on his cars; his clothing was without exterior labels; and his charitable contributions anonymous. Frank V. Lieberman was an expert keeper of secrets.

Whenever I traveled, my father appeared in case there was something I needed at the last moment, which I always did. After watching over my domain in my absence, I would come home to a weed-free garden, a groomed garage, a painted room, a repaired chair, a patched wall. My father would be waiting, wordlessly. My homecoming home task was simple, discovering the differences he had made.

Olivia predicted the self-esteem of many handymen would be rising. Maybe, but I have yet to feel disciplined, contained, competent, or appropriate. Frank

V. Lieberman set hopelessly high hurdles; I will be ducking under them until I die.

A few weeks before his death, my father gave Eben a lavish compliment, "I really love watching your games, Eben. You are such a good all-round player. You are fast, think ahead, and play both excellent defense and offense."

Unequivocal praise from my father was so rare, my face flushed hearing these words. During the fifty-six years our lives overlapped, I could count on one hand the number of times he praised me. During my childhood, we sat across from each other playing hundreds of games of checkers and long chess matches, but Frank V. Lieberman offered no parental lenience; I never won a single game.

Except for his generous advocacy of the CyberKnife treatment, my father never spoke directly to me about my illness. The psychological distance between us was rarely optimal for me, but for him it was learned behavior. Keeping his distance and avoiding conflict was how he survived his childhood. He used those same skills to keep his marital vows, never abandoning his wife or his only child.

My father left no will. My mother's last testament, so full of commands, admonitions, and threats, led him to put everything in joint tenancy with me so I would avoid the long expensive probated torture created by his wife. Instead, all I was left with was my father's last "to do" list—seven chores, four of which he hadn't finished, written on a yellow post-it note.

Last To Do List of Frank V. Lieberman, April 1998

Should I make a quilt of my father's rags to comfort me? His rag box seemed more personal than his wardrobe. It was full of his Hanes undershirts, twenty-year-old Kmart kitchen towels, Eben's tattered cotton diapers, and his own stained linen handkerchiefs. I wanted to stitch a quilt of them, smelling all the scents and sensibility, before wrapping myself in a final embrace.

In the last minutes of his life, I told my father that I hoped he would be surprised to find his disbelief was erroneous. I even tried to assure him that if there is a judgmental god, he had no worries. He was a better Jew, Catholic, Mormon, Protestant, and Muslim than many who have been baptized, circumcised, or otherwise elevated. If there is a loving God, then my father surely was embraced for his fidelity. If there is nothing, his legacy of devotion and hard work will live on in his granddaughter and grandson.

Four months after his death, we buried his ashes in one of the two plots Bob had given me in Columbia Pioneer Cemetery for what we thought would be my last Christmas in 1989. Marley took the lead as we walked the five blocks from our dream house to the cemetery, where two Crist Mortuary morticians and three metal folding chairs awaited us.

Frank V. Lieberman's Gravestone, Columbia Pioneer Cemetery, Boulder, Colorado

Since then on our frequent visits to the cemetery, we place a small rock at the base of my father's gravestone in keeping with the Judaic tradition of honoring the dead. There are other Jewish burials in the cemetery, but we seem to be the only mourners practicing this tradition. Sadly, the cemetery caretakers periodically remove all the stones when cleaning the grounds, without understanding their sacred significance. Were he in charge of cemetery maintenance, my father would likely have done the same.

-44-

Life on Dialysis with Osama Bin Laden

The side effects of cancer treatment are cumulative, like useless credits in the college of existence. While Herceptin combined with Docetaxel was a targeted miracle, the treatments eventually caused kidney failure by accelerating the calcification of my renal arteries. Nonetheless, I consider Herceptin to be a wonder drug. It not only kept me alive, but made me strong enough to witness Olivia's marriage. Like her mother, she married a wonderful Jewish man from Brooklyn.

Survival allowed me to stroke the hair and hands of my father as he died, and to hold my first grandchild in my arms three months later. My Herceptin luck continued when I was able to do the same with Eve, my second granddaughter, a few weeks after the attacks on September 11, 2001.

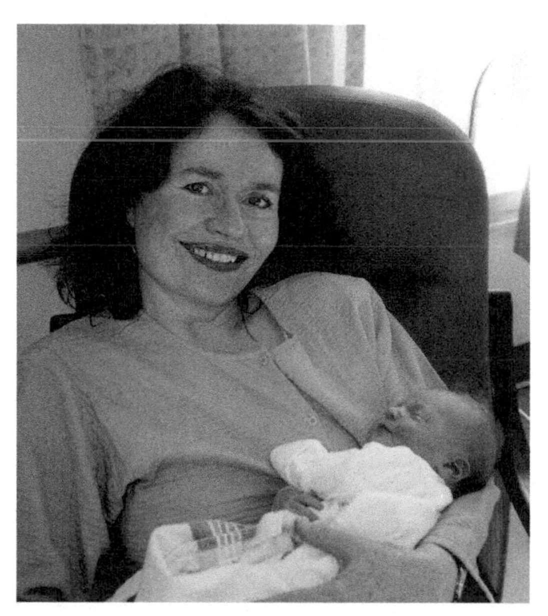

The author holding newborn granddaughter, Eve, New York City, October 3, 2001

Nine months after those attacks, I wept as I watched Eben graduate from Boulder High School as I sat on bleachers in the June sun, with Marley on one side, Bob and Jordan on the other.

Eben Valentine and Marley, Boulder High Graduation Day, June 2002

By September 2002 my body was tethered three times a week to a dialysis machine. Through the din of those long afternoon dialysis sessions, my mind was desperate to escape the sound of FOX broadcasters on the communal television at the Da Vita Dialysis Center, where the manhunt for Osama bin Laden was repetitive news. It had long been rumored that Osama was in need of dialysis. As a distraction from the dying going on around me, I wrote him a letter.

March 5, 2003

Mr. Osama bin Laden
c/o Al Jazeera Television Network
Doha, Qatar

Dear Mr. bin Laden:

Although we have never met and never will, I am writing because I can't stop thinking about you, particularly about our fundamental differences and inexplicable similarities. You are a man; I am a woman. You are very tall; when standing straight, I can claim five feet and two inches. You are an Arab and believe yourself to be a devout Sunni Muslim. I am an American and an agnostic. These ethnic-religious inheritances account for our most important distinction: you suffer from a surfeit of certainty about your religious beliefs, while I suffer from an equal amount of uncertainty.

One of our common attributes is that we are of mixed blood. Your father was from Yemen; your mother was Syrian, with an apostate dislike of veils. You were born in Saudi Arabia and raised in the Islamic faith.

I inherited the genes, but not the faiths, of a confusing mix of people who immigrated to America because of their dangerously unpopular religious beliefs. My paternal grandparents were a Jew and an Irish Catholic. My maternal ancestors left different Protestant sects to join a third, a new one called Mormonism. This was a risky shift. Imagine how it would be if you, a committed Sunni, turned your back on your family to become a Shiite. Actually, it would be more like you becoming a Hindu.

My mother's Mormon ancestors chose to believe the prophecies of Joseph Smith, the latter-day American infidel version of your own Mohammed. Joseph Smith said he heard from his God through an angel named Moroni, similar to the way the angel Gabriel revealed to Mohammed what Allah had on his mind. Pursued and persuaded, the members of my mixed religious diaspora made their way to the state of Utah.

I have read that your mother, Hamidi, being your father Mohammed's tenth or eleventh wife, as well as being an outsider—i.e. Syrian, not Saudi—was called "The Slave" by your father's other wives and children, and you were "Son of Slave." In my case, it was my mother who was considered to have married down. Mormons consider anyone who is a non-Mormon "to be a Gentile" (a Western term for infidel), but particularly one with my father's background. I grew up feeling like a wild animal hiding out among a herd of large dairy cows. Did you feel like a camel in a herd of goats?

I understand that you have never been to America. It is reported you have only traveled as far East as Pakistan and as far West as Sudan, so you may not know that Utah has several geographic and cultural features in common with some of your favorite places on the Arabian Peninsula—a large body of salty water, deserts, mountains, multiple caves, many modestly dressed women, and few out-of-the-way places where polygamy is still practiced.

Utah is a place where the daily lives of its Mormon majority are, like your own and those of your suicidal followers, driven by promises of after-life rewards and upgrades. However, Mormons do not believe in suicide. Plus, this fast-growing sect has nothing as cleverly motivating as seventy dark-eyed virgins. The post-mortem sexual incentives for Mormon men are slightly less misogynistic. The Prophet Joseph Smith revealed that worthy Mormon husbands were entitled to have their wives and children sealed to them for eternity in secret Temple rites.

For Mormon men, however, there is another heavenly incentive tied to the aphrodisiac of power. After death, the most righteous Mormon males will become gods of their own heavenly kingdoms. Of course, even heavenly reward point systems have their hierarchies. Mormon doctrine outlines three reward levels in Heaven. The "Celestial" is the highest degree of exaltation, and consequently is available only to the best people. More importantly, at the Celestial level these Mormon mini-gods can continue to procreate. Sex is not allowed on the second and third levels, the "Terrestial" and "Telestial" kingdoms. The whole hierarchy is similar to the first class, business, and economy seating on our airlines – something with which you are no doubt familiar.

Having grown up in a large polygamous family, you have chosen to continue that tradition. Someone told me that, after your first arranged marriage at seventeen to a Syrian relative of your mother, you, like your father, took three additional wives—four at the same time being the maximum allowed by Islam. Have you also, like your father Mohammed, continued to trade in one wife for a new one every few years?

I inherited some of the impacts of polygamy, but I am three generations removed from its actual practice. Instead, like many other Americans, I have practiced serial monogamy. Unlike you, I inherited neither wealth nor faith from my father, who was a scientist and an atheist. He did, however, leave me priceless memories of his dignity and devotion.

I understand that you and I have begun to share a single preoccupying human vulnerability. Apparently we are both dependent on dialysis. I am curious about how you have managed to survive so long, under what appear to be less than optimal conditions. When I am at the dialysis center, bombarded by the unremitting drone of the pumps sucking and scrubbing my blood, your visage often appears when I close my eyes in search of a semblance of privacy. I believe it is highly unlikely you have been living in a cave all these months, now years. People with your kind of resources and connections usually get much better health care.

During dialysis today, I realized what the CIA should have done after your twin bombings in Africa. It should have started tracing shipments of dialysis tubing, instead of those aluminum tubes. America is so far behind your al-Qaida group in low-tech strategic justice. Had we made it half as hard for you to get dialysis supplies as my insurance company makes life for me, your creatinine levels would have been so high and your blood so toxic, your mind would not have been able to conceive, let alone execute, even a simple act of terror!

I am relatively new to dialysis. After surviving with metastatic breast cancer for many years, the poisons meant to keep the cancer cells in check took their toll on my kidney function. You are only forty seven years old. I have never heard why your kidneys failed, but I am sometimes curious about the origins of your disease.

I am one of the youngest and more mobile patients at the center where I am dialyzed. Most of the other patients are members of what America calls the "Elderly Frail" tribe. They arrive in ambulances or small buses outfitted with motorized lifts, making their way from the curb to the dialysis room with the aid of walkers, wheel chairs, or on a gurney. Inside the center, oversized chairs on platforms are lined up in rows waiting for us as if we were coming for treatment at a large beauty salon. Some patients make it inside with the support of a cane. They make me think of you holding your over-sized shepherd staff, negotiating your way up and down those mountains by your Tora Bora cave.

In addition to kidney failure, you and I have a facial feature in common, which I mention to highlight an important difference in our separate cultural contexts. At the dialysis center, another patient, a gentleman of eighty or more, has taken a fancy to me. He has made his interest known in the way many elderly people do by loudly stating his unconscious desires.

His commentary is crude: "Your lips are very large! I have not been near a woman with such luscious large lips for many years. You weren't here last time I came, but I could still smell your perfume."

We have a saying in America: "He is a dirty old man." This is not a reference to a man living in a cave, but rather to the lecherous intentions on the part of a male too old to be shopping in the jeune fille department. I have been surprised to find myself more offended at age sixty by a dirty old man, age eighty, than I was at twenty by men over forty.

Actually my lips aren't so large; it is just that the lips of most women in his age group seem to have completely disappeared. You, on the other hand, really do have a very full and sensual mouth. Here in America, were we both on the same payroll, working for the same ends, what I just wrote would constitute sexual harassment and I could lose my job. In your fundamentalist Muslim culture, it is more likely I would be stoned to death for being so brazen.

I doubt even you foresaw how the exceptionally creative, grandiose, low-tech, vicious, and destructive strategies to exercise your religiously-driven influence on our world would draw the attention of an American woman, well past middle age, living in the shadows of the Rocky Mountains. But I have become preoccupied with your essence, your existence. You have taken over the space in my brain formerly occupied by the Tylenol Terrorist. This early American terrorist was never caught, after murdering seven random victims with cyanide by secretly tampering with bottles of Tylenol capsules sold in the Chicago area in 1982, only a year after you finished your degree in Public Administration at King Abdul-Aziz University in Jeddah.

Was this early American terrorist the source of your inspiration for your current attempts to manage, menace, and murder the American public, not to mention the populace of Spain, Indonesia, etc.? There were other Tylenol Terrorist copycats; most were greedy disgruntled spouses trying to collect insurance money. Each managed to put poison in multiple bottles of Excedrin or Anacin, knowing his spouse would soon be purchasing that specific brand of pain relief. In each instance, innocent shoppers lost their lives.

Since then, a large number of products sold in my country have been packaged in ways that put the efforts of our new (really it is your new) Homeland Security Department to shame. For years, I have been silently cursing the Tylenol Terrorist each time I am forced to use a box cutter (or pliers, hammer, or screwdriver) to open the everyday products I buy for my family. The sheer waste of this safety program drives me crazy—the unrecyclable protective plastic, the higher product prices required by these pseudo-protections, the strain on my increasingly arthritic finger joints, and the reminder that the sense of safety I once felt is gone forever. Now, instead of the Tylenol Terrorist, I think of you, Osama.

You may object to my addressing you with first name intimacy, but it is a small demeaning cross-cultural price for our physical synchronicity, for your daily incursions into the territory of my thoughts, and for how close you came to hurting my progeny, who live in your favorite target-rich environment. A few weeks after September 11, 2001, my second grandchild was born in New York City. In anticipation of her arrival, I flew to La Guardia the first day commercial air traffic resumed. I wanted to be with my daughter, to comfort her, and help care for the new baby, as well as my first born granddaughter, Tate. My daughter and her family then lived in a high-rise apartment building on East Fifty-Fourth Street, just off First Avenue, an area referred to by locals as Midtown. You can pin point the cross streets on a map of Manhattan, something I assume you have close at hand.

In our collective shock and denial during the weeks immediately after September 11, 2001, Midtown somehow felt a world away from the still smoldering remains of the World Trade Center. The distance was exaggerated by protective levels of progesterone, as well as our well-intentioned parenting and grand-parenting of Tate, who had just celebrated her third birthday. Her parents and I, guardians of her child-like certainty of belief, judged Tate too young to lose her loving trust in our world. So, even though you were in our minds, we adults chose to act as if you did not exist, and pretended that you had not imploded our normative sense of security.

In my role as temporary nanny and baby nurse that fall, I sometimes slept at a one-star Best Western hotel on the Upper West Side of Manhattan. After Tate was asleep and my son-in-law returned from work, I would discreetly disappear to my absolutely awful, but free, hotel room. It was free because I was the victim of Best Western's Rewards Points earned during my husband's frequent business travel. So, every night during the last week of September and the first three of October 2001, I walked across town from First Avenue past Broadway to Ninth Avenue. The sidewalks and streets were strangely quiet, devoid of both people and motorized vehicles. The normal noisy humans of New York City were in mournful anxious retreat. The glamorous restaurants I passed around Rockefeller Center were painfully empty, serving only a handful of people at one or two tables. Excess wait staff, like shocked sentries, stood staring at nothing.

The closer I got to Eighth Avenue, the stronger the scent of the smoldering ruins of the World Trade Center. My olfactory memory of this scent is still evoked every time I think of you, Osama. Your perfume, like my own, is memorable, but yours is not pleasant. It is the odor of raw pain from mutilated and burned human bodies, mixed with the invisible toxicity of melted and pulverized chemicals—your peculiar pheromone. Each time you wander into my mind, your scent arrives too. It is as if I am being forced to sniff a soiled T-shirt you have worn for many months of pseudo-heroic caveman living. My nose is the ultimate detection system. I guess you and I would never be genetically matched, even if we had healthier similarities and more tolerable differences.

Osama, I can imagine that during your childhood, there were far too many times when your existence was barely noticed. It must not have been easy on your ego to be one of fifty or more children and to be called "Son of Slave" by your siblings. Also, your father's death in a plane crash when you were only eleven must have been extremely painful and confusing, coming at a time when you were just beginning to individuate.

This is another of our differences and similarities. While you were the only child of your mother, and lived alone with her in the house your father provided for the two of you, your contact with your domineering father almost always included a large number of your siblings. I also was an only child, but what I have learned about sibling rivalry comes primarily from observations of my two children. My daughter was born when I was twenty-one; my son when I was forty-one. Shortly before her brother was born, my daughter was home from college on vacation and dreamt she was holding a baby, but accidentally dropped him. For most of the next twenty years, during my daughter's semi-annual visits home, my son managed, with uncanny regularity, to create some kind of crisis. It was his way of making certain that his father and I were acutely aware how difficult it was for him to tolerate even the moderate amount of attention we were devoting to his older sister.

So I can well imagine, Osama, how difficult it has been for you to distinguish yourself from your siblings, as well as the depth of your desperation for paternal attention. All members of our species, regardless of their tribal affiliation or religious beliefs, crave attention.

Whenever I feel particularly confused about our world, I recall something my father taught me: "God, as in God the Father, is someone we human beings created in an effort to make ourselves feel special, loved, and under surveillance."

Some days, I have weak hope that your distorted cravings for influence and power, i.e., your adult manifestations of childhood attention deficits, have been satiated in the aftermath of September 11. I would find it exhausting to be so sought after, particularly when you are not feeling well. In the days leading up to dialysis, I feel increasing fatigued and mentally confused. After dialysis, I feel enervated. The periods of normalcy between treatments are steadily shrinking. Increasingly, I find myself just wanting to crawl into a cave and die.

Whenever I begin to think that you have gone on to those seventy impatient virgins, Al Jazeera announces that you have produced one more videotape of inspirational guidance for your followers and threats for my progeny. If you love death so much, why are you working so hard to stay alive? I know that death never shows up on time. It always arrives either too early or too late, but in your case the ironies make me grit my teeth. Except for one former friend, you are the only other human being that I have wanted to see die. Further, I want you to die before me, despite my suggestions that you might be therapeutically rehabilitated through some kind of adjustment in your goals, as discussed above. Also, the methodology I prefer is that you suffer a long fatal convulsion caused by a toxic overload of the excreta in your blood, rather than being quickly martyred by an expensive missile from an anonymous American drone.

I understand that your body guards, who include several of your sons, have been instructed to kill you immediately if there is any chance someone from the West is about to capture you alive. Do you really have a pact that requires a son to kill you? If so, you may have fathered twenty-five biological children, but, you, Osama, are no parent. I am ignorant of what Mohammed, your father, or Mohammed, your prophet, would have said about this plan, but I think it would be better to stop buying dialysis tubing or to acquire your own cyanide, than to ask your child to kill you. It is beyond cruelty for you to leave any one of your children with such an inescapable emotional burden.

During our now long-ago courtship, my husband led me on by teasing me about my "hubris" – a vocabulary word with which I was then unfamiliar. Having since learned what hubris means, I have concluded that another of our commonalties is that, despite our modest protestations, we are both a bit top lofty. Let me begin by giving you a descriptive example of how I manifest presumptive pride.

In the weeks after your destructive assault on our sense of comfort, Tate rode her two-wheel bike, sans training wheels, with unusually precocious three-year-old coordination and agility from her high-rise home on East Fifty-Fourth Street all the way to her preschool on East Seventy-Third Street. Upon arrival, Tate parked her two-wheel bike among the gaggle of strollers in which her peers had been pushed. I was Grandma Hubris!

Walking, sometimes running, beside and behind Tate as she rode her bike to and from preschool, I took considerable progeny-type pride in the large motor skills of my granddaughter, delighted by the "shock and awe" visible in the faces of the developmentally astute pedestrians we passed. Each block of her travels, I found myself imbuing my first-born grandchild with a kind of invincibility. If you attacked again I thought Tate would be able to ride out of danger. This is how I manifest hubris.

Which brings me to yours. I first recognized it while watching a video-tape of you talking with collaborators about the unexpected success your operatives had achieved by having, not just damaged, but demolishing the

Twin Towers. You were proud of their achievements, but you also needed to point out that only you had foreseen the engineering vulnerabilities of the Towers. You explained how you had made calculations that the intense heat from the fire created by the airline fuel would weaken the steel supports. I watched as you gestured with your long elegant fingers (my own are stubby and short, another of our differences), smiling as your audience listened to your words with visible reverence. In that replayed moment, you became a mere mortal like myself. I saw you glowing with hubris. It was if you believed that you had surprised even Allah with your clever plan.

On September 11, 2003, I was once again in New York City with Tate and Eve. After their bedtime baths, they led me up the stairs to the highest point of their new home, a garden on the roof of their condominium building. The moon was full, the sky was cloudless, and a gentle breeze occasionally moved their freshly washed hair across my face as they snuggled on my lap. This is what constitutes a garden of heavenly delights for a Gentile grandmother. Together we watched a full moon play a wondrous game of peek-a-boo between skyscrapers over the East River.

Yet, sadly, you were there with us, Osama. Each time the sound of an airliner or helicopter announced its presence in the sky overhead, Eve raised her hands to cover her head and duck. Her parents have no explanation for this defensive behavior, something she has exhibited from the time she was thirteen months old. Her pediatrician has assured her parents that she will outgrow this response to the sound of an aircraft overhead. Since before her birth, her parents have continued to pretend that you do not exist. Eve did not learn about your having turned airliners into missiles from her mostly pre-verbal friends or from television.

Watching Eve, my aching heart led my agnostic mind into other faiths—even into Dr. Ian Stevenson's rigorously documented cases of reincarnation.[51] *Dr. Stevenson might investigate whether Eve, (who initially spoke with a Long Island accent), could be carrying the worries and trauma of someone killed on September 11. In the end, however, I return to what I know from my own experience: our children carry our burdens even when we try to hide them.*

Another lesson I have learned from forty years of parenting experience is that I should have no expectation about having strategic influence over the behavior of my offspring. Now that you have announced your paternity as the Father, with role-implied control, of many more than fifty terrorists, you appear to be very powerful and have become more famous than your father or any of your many brothers. I imagine the Saudi Royals may now be regretting their insulting decision to turn down your offer to bring thirty thousand jihad fighting followers from Afghanistan to defeat Saddam Hussein following his invasion of Kuwait. During dialysis, I have wondered whether your older, well-connected brothers had a hand in keeping you out of their sandbox or influenced the subsequent decision by the Saudi government to invite Americans to despoil it. Betrayal can only come from those we trust.

The downside risk of trusting your ideological followers is that they will begin to individuate and to pass through their own existential crises. Even though my country has spent billions to develop protective packaging like TSA, and, there is more and more evidence that American leaders have been duped by some banal combination of bad intelligence, bureaucracy, and their own hubris, I still have hope for the future. It seems likely that several of your terrorizing scions will do something stupid. Perhaps one of them will attempt to execute a plan without adequate preparation or, in a moment of inattention, leave an electronic trail to your secluded hide-out.

You have already experienced the defection and betrayal of Sidi Tayyib Al-Madani in 1995. Your choices have betrayed your family—even your mother Hamidi has begged you to stop. You are their familial nightmare. Who will be yours? I have always wondered whether September 11 had something to do with the fact that both your father and your eldest half-

brother, Salem, died in plane crashes. Do you have nightmares about your father's death?[52]

I may be a naïve old woman, but if I were President, I would put more energy into spreading misinformation and increasing sibling rivalry in your followers. Maybe soon an al- Qaida cell will stage a revolt, challenging your too-distant paternal authority, deciding they want your power and fame. Then you will be their easy assassination target.

Still another possibility is that one of your biological sons will walk away from Islam. I can imagine him announcing his defection on Al Jazeera television, telling the world that he is going to America to play basketball for Duke University. Did Mohammed forget to mention, or the angel Gabriel forget to pass on, this important revelation: Childhood development destroys any parental hubris. Whatever we teach our children often comes back to haunt us.[*]

Always thinking of you,
J. C. Lieberman
Somewhere outside of Utah in the American West

* Osama bin Laden and his son Khaled bin Laden (born to his fourth wife Siham Sabar in 1988) were killed by American Special Forces on May 2, 2011 in Abbottabad, Pakistan.

-45-

Marley Upside Down

After my father's death, Marley took the lead whenever we retraced his favorite trails. She always looked as if she expected to find him just around the next bend. Sadly most of their favorite hikes were beyond my diminished physical capacity. In an act of loving kindness, Charlotte Smokler began taking Marley for a walk on Sunday mornings. Marley knew whenever Charlotte was due, always waiting expectantly at the top of our stairs.

When Eben started college in 2002, our house grew increasingly quiet. Marley was slowly losing both her vision and her hearing, having twice undergone surgery to remove cancerous tumors. We began pushing Marley in a stroller on walks across the campus of CU. Over the next year, as Marley's quality of life diminished and her pain became harder to treat, we knew our time with her was over.

I took Marley for a last visit with Professor Lawrence Senesh. Professor Senesh, a widower since Dorothy's death in 1989, had long suffered from Parkinson's disease. I violated the condominium rules by taking Marley up to his unit so the two of them could have a sweet reunion.

As we were leaving Professor Senesh said, "Marley, I will see you again soon and we will be together forever."*

On September 4, 2003, Marley's veterinarian, Dr. Catherine Eppinger, came to our home. A few hours before she was due to arrive, Bob dug a grave for Marley in our rear garden on the slope of Flagstaff Mountain.

Dr. Eppinger administered the fatal injections as Marley lay on her favorite blanket in the late summer sun. I was inchoate as Bob wrapped Marley's blanket around her body before he struggled to carry her dead weight two hundred feet up the hill. Bob was at an awkward angle when he lay Marley's body in the grave

---

* Lawrence Senesh died a few months later on November 19, 2003.

and as a result her head ended up at the lower end. I could hardly sleep that night imagining Marley's discomfort.

In death, Marley on her favorite blanket, Boulder, September 4, 2003

For many days after I felt an urge to excavate Marley's body in order to place her head at the higher end of her grave. Several weeks passed. While I was still catching imaginary glimpses of Marley in her usual places, I eventually was able to visit her grave without weeping—although not without wanting to make her more comfortable. Then one night, I dreamt my father was derisively making fun of my concern and scolding me.

"Stop your fussing, Mama! Marley is dead! There is no pain or discomfort after death!" my father said as tears streamed down his face.

-46-

Saved by Switched Seats

Dialysis is a peculiar form of life-support. Even the American Kidney Association offers advice to patients on how to stop treatment when someone on dialysis has had enough. After three years on Herceptin, and before becoming dependent on dialysis, I took fourteen months of treatment with 5-FU, still believed to prevent a terrible end in the Church of Chemotherapy. A year later I joined a trial for a new version of Taxotere (Docetaxel). The last clinical trial I entered was in 2001. It was one of the first vaccine trials for immunotherapy, the latest weapon to be hopefully-hyped in the War on Cancer.

In between, I was given injections of mistletoe sent from Switzerland, each time thinking of Nonny and crossing my fingers that it was aborting my tumors.

On the moderately good days, in between dialysis, I continued researching my father's family history, as well as that of Bob's family.[53] In 2006, a group of genealogists put out a call for papers to be presented at their 2007 conference in Salt Lake City. Surprising myself, I submitted a proposal about the meaning of *optional distance* for emigrants leaving their European families forever to build new lives in America. It was accepted by the conference committee, leaving me with several months to prepare.

Those months flew by. Jordan fell in love with a wonderful girl named Florence, who accepted his proposal of marriage. I developed a raging infection in my jaw that required three days in the ICU. After that my beloved Aunt Mary died,[54] and cousins, children, and grandchildren came for visits. In May 2007, my nephrologist discovered I had developed an abdominal aneurysm. Having already decided against any further surgery or interventional cardiology, the only option was medical management of the condition. After a third ultra-sound revealed an increase in the diameter of the aneurysm, a substitute nephrologist called in an arrogant young surgeon.

He looked at the images and announced, "I am going to admit you tonight for surgery in the morning."

"No," I said, "I'm going home."

That night I dreamed I was moving into a new house with a large window extending from the first floor to the basement. The window was similar to the two-story green-leaded glass window in the dining area of our Marine Street home, a window that had come from an old church and provided light to the basement hall below. In the dream, I went down into the basement and was surprised to find the area was sunny and light. A few weeks later, I dreamed I was falling down a long dark flight of stairs with Aunt Mary. We were holding on to each other as we tumbled down. I had the same dream three times, like a recurring fever dream, but each time I awakened before we stopped falling. The stair dream came almost a year to the day that Aunt Mary, age ninety-six and confined to a wheelchair from crippling arthritis in her feet, got up in the middle of the night, opened the door to her basement stairs, and tumbled down to her death on the cement basement floor. My most recent trip to Utah had been to attend her funeral.

Olivia had a hard time accepting my decision to forego surgery and mobilized her Manhattan connections. A referral was made to a more experienced vascular surgeon in Denver. He and I only had a virtual consult. After a long day of surgery, he was hooked up to a live feed of my fourth abdominal ultra sound. He directed the radiologist's wand speaking about what he saw as if I was in another room, understandably annoyed by the need to fit me into his busy schedule at the request of an important hospital donor. He said the risk of surgery was high.

Over the next week, I began to feel strangely incapable of making even simple decisions, a side effect I attributed to the zombie-like-effect Beta-blocker drugs have on my persona. The presentation at the Salt Lake conference loomed, but I now found it difficult to imagine myself traveling to Utah by any mode, let alone negotiating my way through an unknown dialysis center or following my notes on the dais. The conference co-chairs were notified of my need to cancel. They were gracious, but persistent. They offered special assistance to accommodate my needs, telling me I could come for only one day instead of the full conference. Their pleas triggered my receptors for being over-responsive; I told them I would think about it overnight.

I was also working against a different deadline, struggling to design and print invitations for a wedding picnic for Jordan and Florence. I was overwhelmed by slowed thinking, complicated by the challenging relationship my generation has with technology.

When I took the last invitations to the post office, my path coincidentally crossed that of Ginny Wells Jordan, who had first introduced me to the Gonzalez therapy. We sat down in a nearby coffee shop, ordered tea, and tried to catch up on a decade-sized gap, savoring our mutual survival. Afterwards, my spirits elevated by a single shot of socialization, I decided I could manage a short luggage-free trip.

Federal Express picked up my presentation materials and suitcase the day before my departure. A driver was scheduled to meet me at the Salt Lake City airport who would take me to the Hilton, where I would be guided through an orientation to the audio-visual system. Then there would be time for a nap in my room. I even imagined myself buying one of Utah's miniature liquor bottles for a celebratory drink after my presentation. Defensively medicated by alcohol, I could leisurely stroll around Temple Square before flying home free of the weight of Gentile guilt.

Early the next morning I climbed aboard the bus from Boulder to Denver International Airport. Stepping off the bus, I felt lighter, more like myself than I had in weeks—the gift of vacating ordinary life. Because I had a pre-assigned aisle seat and no luggage, I lingered in the gate area watching the baggage carts and other tarmac traffic until the last boarding call. It was a small plane, with only two seats on either side of the single aisle. As the last passenger to make my way through the fully occupied cabin, I was surprised to find my assigned aisle seat occupied by a young mother, holding an extremely unhappy infant. The man seated next to her was holding the baby's bottle. Before I could open my mouth, the young mother spoke over the baby's screams.

"I hope you don't mind my taking your seat!" Gesturing to her seat mate, she added, "My husband and I couldn't get seats together, but I need his help with the baby! My seat is just two rows further back, by the window."

"Okay, I understand," I said, proceeding to her row.

The stewardess was urging me to take a seat so she could make her safety presentation, but a large young man in the aisle seat didn't stand up when I pointed to the window seat. So I squeezed my buttocks across his knees at a forty-five-degree angle. He was from somewhere that started with O, maybe Omaha or Oklahoma. The plane seemed small; my seatmate was huge. He had stowed a bag on the floor beneath the seat in front of mine. He offered to move it, but I instead wedged my black tote under my seat beneath my knees.

The two of us had a brief conversation. Mostly he talked and I listened. He was a paramedic firefighter on his way to a wildfire in Idaho. After telling me he would be joining a "hotshot" team, he confessed that this was his first time on an airplane. His excitement and enthusiasm for his new adventure was palpable.

"You smell so great!" he said, not once, but three times, failing to notice that I was not flattered.

Clearly, he had an excess of hormones, where I had none; also, I was old enough to be his grandmother. When there was a pause in the conversation, I must have dozed off with a cup of coffee in my hand.

The next thing I remember is the impatient voice of a stewardess, saying, "Excuse me, Ma.am, I need to collect your container!" Can you please wake up?

Opening my eyes, I could see nothing as my still-full cup of coffee fell to the floor.

Pressing my shoulder against the paramedic, I attempted to say, "Sa so me ink is rong! Nah see ony ting!" (Something is wrong! I can't see anything!)

My enunciation suggested that the stewardess had served me several tiny bottles of liquor, rather than a single cup of coffee. Miraculously, because of his hotshot training, the paramedic spoke my body's language. My sudden loss of vision and slurred speech convinced him I was having a stroke. He took charge like a bloodhound hitting a target scent.[55]

Lifting me out of my seat, he laid me on my back in the aisle, with my head oriented toward the rear of the plane. I heard him commanding the stewardess to page for any doctor on board, while instructing the passengers to remain in their seats. The paramedic continued issuing orders to the stewardess, ignoring my garbled attempts to tell him I needed to use the lavatory.

"Tell the pilot to land the plane away from the gate area ASAP and have him order an ambulance to meet the plane! The patient will disembark at the rear exit!" he said, now sounding considerably older than an eighth-grader.

I vaguely remember him giving me an injection as the plane ground to a sudden stop. Soon two ambulance attendants began struggling to load me onto a gurney. As they lifted me off the floor, I realized it was too late—I had wet myself. The ambulance men maneuvered the gurney out the back door. About half way down the stair ramp, I began to vomit.

My personal paramedic shouted instructions to the ambulance attendants from the top of the stairs, "Turn her over! Now! Before she aspirates!"

As the two young men attempted to follow the paramedic's commands, my body slid forward, apparently causing one of the attendants to lose both his balance and his grip. He let go of the gurney as he jumped safely onto the ground. Then it was as if I was back in my stair dream, but without Aunt Mary holding me in her arms. I slid off the gurney, my head leading the way to the tarmac. I later learned my fall was estimated to be about six feet, but it was not the best moment to be playing slip and slide, lubricated by vomit.

For the next eight hours I was either unconscious or anesthetized. I was taken to the emergency room at the University of Utah Hospital—a Trauma I facility—an "unknown patient, female, estimated age fifty plus." Separated from any identification, as well as my normal communication structure, I arrived incognito. When the plane came to a sudden stop, my tote bag remained wedged under my seat out of sight. After the plane went on to Seattle, my "unattended baggage" was a stowaway on its own adventure, mysteriously missed by those meticulous cleaning crews and by two sets of supposedly alert passengers.

Eventually the airline provided the hospital with the name, address, and phone number of the female passenger assigned Seat 17A. A hospital official called her home. Her husband answered.

"Was your wife Christy, scheduled to arrive this morning from Denver?" the official asked.

"Yes," he said, "She is already home."

Ignoring the seemingly incongruent information, the official informed Christy's husband his wife was in critical condition and asked him to come to the Hospital Emergency Room immediately.

The husband responded, "You must have the wrong number. My wife is home safe and asleep."

The hospital official asked to speak with her.

"No, I don't want to wake her or the baby," the husband said and hung up.

A uniformed police officer was dispatched to Christy's home. Her husband answered the door.

The officer queried, "Are you certain your wife has returned? Could you please ask her to come to the door?"

When told no, the officer tried, "Could you give me a physical description of your wife?"

An increasingly annoyed husband responded, "My wife is five feet, eight inches tall, she weighs one hundred forty pounds, and has blond hair and hazel eyes."

Policeman: "What is her age and birth date?"

Husband: "She turned twenty nine last week, on the twelfth."

Policeman: "Was she traveling with anyone?"

Husband: "Just our six-month old son."

Christy's husband had no knowledge his wife had switched seats because he had not been on the flight. He was not the "husband" sitting next to Christy, helping with the baby. An adulterous secret apparently contributed to my loss of identity. Had I been identified at that point, I believe Bob would have carried out my advanced directive against any further surgery and I would have died in Utah, not far from where I was born.

My accidental slip and slide had fractured my skull along my right orbital bone; an epidural hemorrhage was in progress. An MRI was done of my head and the paramedic's assessment of vision loss due to stroke was accepted. The statements of the ambulance crew as to the cause and the nature of my blunt brain trauma injuries were recorded.

I was lucky a fourth time because it was ten o'clock in the morning on Friday, not ten o'clock on Friday night, or any time on Saturday or Sunday. I had arrived on a day and at a time when an experienced surgical team was on site. The young paramedic had impulsively given me a shot of blood thinner. In light of my abdominal aneurysm, I had luck a fifth time because the profuse bleeding caused by the fall kept my blood pressure so low, the aneurysm couldn't have ruptured even if it had received a direct revelation from God via Joseph Smith. Utah has always claimed an excessive number of miracles, but these events added up to so much coincidental luck I was subsequently forced to question my agnostic ethos.

Unlike Humpty Dumpty, the surgical team miraculously put me back together again. After drilling several holes in my skull, they used an endoscope to stop the bleeding from a tear in the temporal artery. Eventually, they performed a double endarterectomy, i.e. they sequentially opened up my carotid arteries, cutting out gobs of butter covered in dark chocolate. The source of the stroke was a clot that had formed in my left carotid, but both carotids were more than ninety percent occluded.

On a genetic side note, my father's family tree is loaded with hyperlipidemia victims. They also had a proclivity for developing abdominal aneurysms. His brother died of a dissected one. My father's aneurysm was discovered when he was exactly my age, sixty-five. His vascular surgeon felt he should have his carotid arteries cleared before proceeding with the abdominal repair, which he did.

Because I had plates with screws covering the front and back of my cervical vertebrae, it was necessary to open my chin to get adequate access to the carotids. The surgical team then began to slowly cauterize their way out from my eye sockets to the top of my skull, stopping the bleeding as they stitched the skin covering my brow and my eyes to my former hairline, closing multiple incisions on my skull. I was alive, but unrecognizable.

My first moment of consciousness came about three o'clock in the morning, almost nine hours after the surgery ended. In the recovery room, as someone was attempting to suction blood from my nose, a surgical team member asked, "Do you live in the Salt Lake area? Were you coming to visit someone?"

All I could muster was a slurred Yiddish-sounding version of "Yewish yenes."(Jewish genealogy).

Someone on the team had read about the conference in the *Salt Lake Tribune* and called the Hilton Hotel. A conference coordinator was awakened; she already knew who I was—the presenter from Colorado. The driver she had sent to the airport had waited three hours with my name on a large piece of cardboard. My mysterious absence had created a big scheduling mess for her to manage. Her name was Miriam and she had planned to call me to try to find out why I had failed to appear, but she had too much to do with the closing banquet and the chaotic departure needs of conference participants from a dozen different countries.

Miriam saved me. I could barely see her, but her gestalt was clearly that of someone used to running central command. Her voice was authoritative and demanding. I subsequently learned she was the widow of a rabbi and the mother of six children; hence her well-honed executive abilities.

Miriam held my hand: "Are you Joan Lieberman?"

I nodded "yes" and incoherently began begging, "Pleese! Yelp me! No vand die in Vtah. Go lome to tuban?" (Please! Help me! I don't want to die in Utah. Can you help me go home to my husband?).

Her response was instantaneous, "I understand completely. I will get you home. May I call your family to make those arrangements?" (The only number she had was for my cell phone. My tote bag and my cell phone were still on the aircraft at the Seattle airport.)

I tried to shake my head "No" but it seemed immune to my command. I tried to say, "Vill frighten. Wand to aite." (Will frighten him. Want to wait).

"Alright," Miriam said. "We will take it one step at a time."

A hospital lawyer was called to my bedside. I was asked questions to which I nodded assent empowering Miriam to act on my behalf. How she arranged this deserves a Gold Medal. Unrelated women are not supposed to have such rights in hospital settings. When the surgical team made their rounds, I heard Miriam arguing with them about my desire to die at home.

"She has every right to make this decision!" Miriam announced in a voice that echoed toward me across overly waxed floors.

Three different doctors, the hospital chaplain, a too-sticky-sweet social worker, and finally a psychiatrist were brought to my bedside before sunrise to certify that I was competent to sign myself out of the hospital against medical advice. I became more adept as the certification process proceeded. By the time the psychiatrist showed up, dressed in a white short-sleeved shirt and a black tie as if he was a Mormon missionary, my alertness was on the rise.

I impulsively attempted to enunciate, "I want to be discharged because I don't want to die in Utah," my syntax sounding like I was speaking inept pig-Latin.

As soon as the garbled words were out of my mouth, I knew I had made a mistake. Given my medical condition and the cultural context, I was talking like a crazy person.

"Do you understand why in the opinion of the surgical staff you need to remain here so your condition, which even you seem to know is critical, can be monitored?" the psychiatrist asked.

"Yaes," I said, still speaking in tongues and thinking of Pepa, my stroke-stricken grandfather, as I made a feeble attempt to say: "I have been told it is unlikely I am going to survive if I try to move, but nonetheless it is critical that I remain at a ninety degree angle for the next forty-eight hours with ice bags on my head. Can you think of a better place to insure that I remain in this position than on an airliner?"

In order to translate my garbled speech, the psychiatrist must have had to use Joseph Smith's imaginary ancient Israelite seer stones, Urim and Thummim.[56] Joseph Smith claimed Moroni thoughtfully buried those seer stones along with the golden plates to help him translate hieroglyphic text.

While I could barely see the face of the psychiatrist, he turned toward the head surgeon, and what apparently was a group of psychiatric interns, to pronounce his diagnosis.

"Her cognitive abilities are clearly intact," he said.

Just after eight in the morning, I was positioned on a gurney at a ninety-degree angle and taken by ambulance to the Salt Lake airport. In retrospect, I think Miriam should have been the lead negotiator in the Israeli-Palestinian peace talks. She had managed to get the ambulance company to provide a private air ambulance to Denver. Somehow, with Miriam's wise and powerful ways, my unaccompanied tote bag and cell phone were located and flown back to Salt Lake.

When the back door of the ambulance was opened, Miriam was there, carrying my purse and a shopping bag with what was left of the clothes I had worn on the plane. Having recovered my clothes from the trauma room, she had taken what was salvageable to the Hilton laundry. She explained what was going to happen and handed me my cell phone. She told me she had listened to my messages and that my husband was a very worried man. I will always be grateful to the inventor of the auto dial function on cell phones. I couldn't seem to remember my husband's name until I heard Bob's voice.

A chunk of grief rose up my throat, "No alk, kooming jome." (I can't talk right now, but I am coming home!)

Bob later said my strange speech led him to think I had overdosed on something. Still wearing hospital garb, my head tethered to dangling drains and ice bags, I was loaded onto an air ambulance, well-secured to another gurney with a fancy wedge pillow followed by two IV bags which were infusing fluid, morphine, a steroid, and two antibiotics. Hooked up to a new monitor, leads and oxygen, I was a scary sight. I looked as if I had been attacked by a vicious grizzly bear. Both bruised and ghostly white, I had four drains coming out of my head, two on top and two behind my ears. My eyes were tiny purple swollen slits with stitches that from a distance vaguely resembled eyelashes. My ears looked as if they had been recently reattached; my neck was purple, as if the grizzly had attempted strangulation.

As soon as we were in the air, I fell into a deep sleep. As one of my attendants awakened me for the landing at DIA, the flight seemed to have taken only minutes. There was tedious transfer to another ambulance on an overheated tarmac, followed by an uncomfortable ride to St. Luke's Hospital. Wheeled into the ER, I was greeted by two "tutt-tutting" ER doctors. Then there was a seemingly endless transfer of papers, prescriptions, permissions, and medical supplies, the disconnection of IV's and monitors, removal of catheter, draining of drains, review of care instructions and medications, more warnings and recommendations.

Finally, one of the nurses helped me into my slacks, leaving my Utah gown in place. What was left of my hair was stiff with blood and Betadine. Miriam had thoughtfully added a prayer shawl to my wardrobe bag and one of the St. Luke's nurses used it to cover my head and neck as if I was going home to Osama bin Laden.

After the doctors made one last attempt to persuade me to move into their facility, a nurse helped me into a wheelchair. With my sack of medical supplies, I was pushed toward a bonded mobility transport and rolled into the van. The ride to Boulder was by far the hardest part of my journey home. When the van finally stopped in front of our home, I almost fell off the wheel chair lift due to an absence of depth perception mixed with desperation to be in my husband's arms. Bob signed the requisite forms and took possession of a woman who only vaguely resembled his wife.

I had received far better care for my injuries than I would have in Afghanistan or Africa. It may have been irrational, but when the doctors told me I was close to death, I felt desperate to say goodbye to Bob. The scene at our house was chaotic. Bob and Eben had no idea what to expect. Supplies were unpacked; extra ice bags and surgical gloves acquired. Through the haze of residual anesthesia and morphine, I overheard them bickering over protocols for maintaining a sterile field. The instructions were complicated. My head was to be held at a ninety-degree angle and kept covered with ice bags for forty-eight hours—challenging requirements in any institution. Fourteen different incisions and four drain holes in my skull were to be cleaned with a fifty percent solution of hydrogen peroxide and then treated with antibiotic ointment every four hours. The caked blood had to be slowly bubbled out of my ear canals.

There were pain pills, antibiotics, steroid tablets, steroid eye drops, cleaning drops, moisturizing salves to administer on a twenty-four-hour clock. There was to be no lifting, no bending over, and no heart rate in excess of eighty.

My desire to see Bob and Eben one last time had been so great I failed to recognize that I was involuntarily drafting my husband and son to run the equivalence of an ICU unit. I have to say my outpatient care was excellent. They provided tender, nitty-gritty nursing without visible revulsion. Bob and Eben, in turn, were supported by our friends and neighbors, the staff of the dialysis center, and Olivia's loving long-distance concern, and with the tolerance and prayers of Bob's clients.

Ten days post-op, I began attempting to regain my independence and to handle my own care. At twenty days post-op, I was able to handle the care of all but four of my incisions and the skull holes were crusting over. At thirty days post-op, I was able to type using the keyboard, although I was not yet able to read the text. There were some very serious cognitive gaps—finding words was not only hard and humbling, but exhausting. My narrative speed slowed to a crawl and any distraction seemed to erase my short-term memory.

My father had taught me that in order to keep living, one has to keep moving. To avoid more clots, I started trying to walk, albeit at half my normal pace. I began to walk the neighborhood before dawn, slowly and carefully, shy of being seen and paradoxically worried I was unrecognizable. Inside our dream house, I felt safe. I could not think about the future. Instead I lived from one four-hour medication cycle to the next, while laughing with Bob about the miraculous calamity in Utah—an experience that created a disturbance in my psyche.

Rainbow after Jordan's rain-soaked marriage ceremony, Denver, September 2, 2007

Two months later, it poured rain as I stood watching Jordan pledge his troth to Florence in a garden ceremony. Immediately afterwards the sky filled with a spectacular rainbow.

Somehow my heart and vascular system had inexplicitly carried me through a major trauma, but dialysis was increasingly depressing. I did not want to be dependent on a machine. My future was long gone; my whole body felt brittle—a badly damaged specimen, lacking numerous necessary parts, preserved by poisons.

Yet, my nephrologist saw me as one of his healthiest patients—just one indicator of the frightening health profiles of most dialysis patients. He had a surgeon soldier friend who was enthusiastic about the improvised replacement parts (IRPs) being used to save the lives of American soldiers blown apart by Improvised Explosive Devices (IEDs). The surgeon soldier recommended replacement of my calcified renal arteries with synthetic ones, the same kind of material that would be used for a by-pass repair of the abdominal aneurism. My nephrologist argued that I had nothing to lose and a great deal to gain. With Bob's support, I decided to take the surgical risk of having a double procedure to repair the abdominal aneurism and replace the renal arteries. Since my carotids had been cleared, it would be like having a full vascular make-over. I repeatedly dreamed of Ben Evans working on an old car in his Logan garage.

The Monday after a quiet Thanksgiving weekend, Bob drove me to the hospital for pre-surgical testing and dialysis. The two-part surgery was scheduled for Tuesday for which there would be two surgical teams, one for the aneurism and another surgeon for the more complex renal artery replacement. It was snowing heavily as Bob started to turn into the parking lot, but I asked him to drop me at the hospital entrance, arguing that he should to get on with his work day because my greatest need for his support would come after the surgery. He protested; I prevailed. If he had parked, it is likely I would have died in the parking lot. As soon as I walked through the automatic doors, I collapsed. The aneurism had burst.

The young surgeon, who was scheduled to assist with repairing the aneurism the next day, happened to be in the hospital. Both of the other lead surgeons were in Denver. Somehow the young surgeon performed both procedures successfully with instructions via video feed from the renal artery specialist.

When it was all over, the young surgeon came to my bedside on Wednesday morning and announced, "I've just been told that because of you I have now experienced the high point of my career. For me, it is all downhill from here!"

It was not all downhill for me. His courageous work created an extension on my life lease of ten years and counting.

-47-

Searching for Ancestral Secrets

What haunts are not the dead, but the gaps left within us by the secrets of others.

Nicholas Abraham*

For Mormons, genealogy is theology. Among the megalomaniacal tenets of Mormonism is the belief that anyone who dies without having been baptized into the Church of Jesus Christ of Latter Day Saints will be unable to advance in heaven. Since billions of people lived and died before the Angel Moroni ever appeared in the visual field of Joseph Smith, Mormons began practicing a ritual known as "baptism for the dead," an earthly act meant to have heavenly consequences. Those rituals began in 1840 and anyone familiar with www.FamilySearch.org or www.Ancestry.com can see how Joseph Smith's vision has been transformed into an invaluable gift for genealogists. Mormons have been collecting and filming vital records from all over the world for almost two centuries in an overly-ambitious effort to baptize by proxy every human who has lived on earth.[57]

But back to genealogy. It continues to surprise me how little each of us know about those who came before us. Few of us develop an interest until we are closer to a date with death than to the date of our birth. Even then, the attitude of many of us with respect to our ancestors is that we were born under conditions resembling an immaculate conception.

When I was eleven and living as a Mormon in Logan, the Mutual Improvement Association (MIA)[58] gave me an assignment to learn about my

---

*"Notes on the Phantom: A Complement to Freud's Metapsychology," Nicholas Abraham, translated by Nicholas T. Rand, page 171. *The Shell and the Kernel, Vol. I,* Chicago, University of Chicago Press, 1994.

ancestors. I came home from the Ward House with an empty pedigree chart and innocently sought help.

"Daddy, what was the name of your daddy?" I asked.

My father turned away from me and left the room. His abandonment by Jacob Liebermann had solidified into historical banishment. My mother glared at me before directing that I was never to raise the topic again. I obeyed until death was on my doorstep.

The 1992 interviews with my father and Aunt Alice gave me the first real insights into my paternal inheritances.* Their stories raised many new questions, most of which the two elderly siblings were unable to answer. After my father's death, I went in search of other sources, hoping to leave more complete ancestral portraits for Olivia and Eben. I spent a week at the LDS Family History Library in Salt Lake City, scrolling through old microfilms on cranky readers. An inexperienced amateur, I returned to Boulder with only one useful Lieberman record.

The upside of that week in Salt Lake City came after library hours when I interviewed Aunt Mary. Over dinner in a Salt Lake restaurant, I began asking Aunt Mary about a few of the Beck family records I had found.

"The marriage record for Pepa and Nanie confused me," I said. "It shows they were married in Blackfoot, Idaho only six months before the birth of Aunt Vermilla. Do you think I found the right record?"

Continuing to butter her piece of stale restaurant bread, Aunt Mary let out a long sigh—"Yes, you found the correct record. But, Vermilla doesn't know, so please don't tell her!"

Laughing at this still-to-be secret pre-marital sex, I replied, "I really wish someone had told me about this a long time ago! I thought I was the first woman in our family to have become pregnant out of wedlock. Instead, now with better sex education and birth control, maybe I will be the last!"

---

* See the chapter entitled "Death, Memories, and Betrayal," page 98; also *An Oral History of the Tuite, Lieberman, and Marshall Families – The memories of Frank V. Lieberman, Alice Lieberman Marshall and Arthur William Marshall, transcripts of interviews with Joan C. Lieberman in Boulder, Colorado*, February 1992. (Unpublished manuscript).

During that same dinner, Aunt Mary told me all she knew about Nanie's care of her sister Genevieve and niece Beulah when they were both critically ill with the Spanish Influenza.

"I was only six, but mother talked about it for years. It broke her heart when Beulah died in her arms. Sadly, Aunt Genevieve never had another child. Over and over she told me that death is a really terrible thing. It seemed like Mother was gone for weeks after Beulah died. Mamie Torkelson took care of us. She was a live-in mother's helper. Neither Mamie nor Dad told us that Mother had caught the virus and had self-quarantined herself at Aunt Genevieve's place trying to keep us four children from catching it too."

Then Mary casually added a critical fact, "It was a very hard time for Nanie because she was pregnant with your mother then."

That dinner table talk with Aunt Mary turned out to be a genealogical gift. It provided a critical clue to the source of my mother's schizophrenia.

It took me another fifteen years of episodic research to piece together the Lieberman tree and almost as long for my father's maternal lineage, the Tuite and Monaghan families. Both trees revealed unstable attributes in religious faith and nomenclature. My father's paternal ancestors were orthodox Jews whose descendants slowly descended into the stark reality of my father's atheism. My paternal grandmother left the punishing doctrines of Catholicism for the less rigid rules of Congregationalism.

After my maternal ancestors switched from Presbyterianism to Mormonism and adopted polygamy, the exogamy of my parents led the way to my early apostasy and excommunication. If you don't know which one of four tribes you belong to, which rules you should follow?

With respect to unstable nomenclature, my paternal ancestors perpetually changed names, from Raphael Levy to Levy Raphael to Fohlen Levy to Leibmann of Metz to Libermann to Libmann to Leibermann or Liebermann. For the most part, Jacob Liebermann and his family dropped one of the "n's" soon after moving to Utah in 1913, although Jacob resumed using that spelling of his surname after he fled to New York. During World War II, my father's oldest brother, Robert, a resident of Chicago, who used "Leiberman" when spelling his surname, came home from basic training to find both a Jewish star and a swastika scrawled on his front door. He tried, but failed, to persuade his two younger brothers to change the family surname to "Leigh"—but both Frank and his brother Edmund had already begun their professional careers. Only Robert changed his surname to Leigh.

Given names were also changed over the last three generations: from Judas to Fallentin to Valentine and from Francis Valentine to F.V. or Frank V. to BaBa. Tecla became Anna, causing her son to begin calling his wife, Nancy, instead of her birth name, Anna. I began my life as Carol, crossed into Montana to become Joan Carol, then J.C., Joan, and finally Nana Joan. Similar evolutions are found among my maternal ancestors.

My father inherited as many family secrets as I did. He never knew that both his parents were born twins. He might have guessed his father was a twin since both he and his sister Alice had been told Jacob Liebermann weighed less than two pounds at birth and had been kept alive in a shoebox on top of the stove. Jacob's twin brother was stillborn, so it is likely that, like me, Jacob Liebermann never was told. Given the era and under those the conditions, Jacob Liebermann's survival was its own kind of miracle.

Jacob's father, listed as "Fallentin Leibermann" in the 1880 U.S. Federal Census, was born in 1820 as Judah Libmann in Alsace. Fallentin was his Jewish temple name, which he began using in America because "Judah" was "too Jewish." Even Fallentin was soon Americanized to Valentine. His first marriage was to Rosa Fuld in Frankfurt am Main. She died during childbirth. Valentine was a fifty-four-year-old widower when he married Jacob's mother, Tecla Levy Goldschmidt, in New York City. Like so many others, Tecla Americanized her name to "Anna Schmidt."

Tecla Levy Goldschmidt and Fallentin Leibermann with daughter Rose, New York City, 1875

Two weeks before her marriage to Valentine, Tecla gave birth to a daughter. Born into an orthodox Jewish family in Bavaria, Tecla was twenty-seven years younger than Valentine. The couple had eight children as fast as biology would permit, including two sets of twins. That birth rate was normative for an era when restraint was the only means of contraception. Nevertheless, since Valentine was sixty-seven and Tecla was forty at the time of the birth of their last child, their fecundity seems to have been on the edge of normative.[59]

Like me, Anna Gertrude Tuite, my father's mother, also had a twin brother, Amico, who survived only thirteen days. Anna's parents, Bernardine Tuite and Bridget Monaghan, were born into Catholic families in County Westmeath, Ireland. Having disembarked from different Irish famine ships in different years, Bernard and Bridget met in Manhattan and married at Saint Columbia Catholic Church in Newark. The couple opened a bakery a few blocks away, where they produced bread and cakes, along with six children, three of whom survived—Robert Parnell, Francis Torquacious, and Anna Gertrude.

Bridget was forty-two and six months pregnant with a second set of twins when she died of preeclampsia in 1899. She was buried the next day with the fetal twins still in her womb. Five months later, Bernardine put a pistol to his head and pulled the trigger. Bernardine's cousin, Frank Tuite, was a Newark police detective, who persuaded the coroner to list the cause of death as a "cerebral hemorrhage" since the sin of suicide prohibited a Catholic funeral mass or burial.

Bernardine's death meant my father's mother was an orphan. She and her brother, Francis Torquacious (a.k.a. Torty) were put on their own version of the "orphan train"[60] and sent to Ogden, Utah to live with their father's sister, Elizabeth Esther Tuite, and her second husband, Edmund Joseph Harris. The childless couple, (who became "Aunt Lizzie and Uncle Ed" to Anna and Torty, as well as to Anna's four children), were non-Mormon immigrants from different British Isles, who owned The American Bakery in Ogden. Those rotting Irish potatoes turned the whole family into worshipers of wheat.[61] There was no formal adoption because Anna and Torty were old enough to understand their orphan status. Anna continued her education in Ogden, but Torty never went back to school; instead he went to work in Uncle Ed's bakery.

As statistically unlikely as the union of my parents was, the marriage of my father's parents broke all the rules during an era when Gentile marriages were even more taboo. Seven years after she arrived in Ogden, Anna Tuite boarded a train and traveled back to New Jersey. She wanted see her older brother Robert and her closest cousin, Charlotte (Lottie) McKenna, whose mother was the daughter of Bernardine's brother. Lottie McKenna's family lived next door to the Liebermann family on Stephens Street in Belleville. (Proximity is destiny in my family).

On the first day of her New Jersey visit, Anna was sitting on the McKenna's front porch with Lottie when Jacob Liebermann came home from work. Jacob, who already had a reputation as a Romeo, was drawn to Anna's slim figure and her beautiful dark hair. Deciding to make a move, Jacob introduced himself.

"Hello, Lottie!" he said, "How are you today? Who is your new friend?"

As soon as Lottie introduced Anna, Jacob said, "Anna is also my mother's American name, but she is not nearly as beautiful as you!"*

Jacob was darkly handsome and his pheromones caught Anna's attention. In that era, she was approaching spinsterhood, and this may have been the moment when she began understating her age by several years, a not uncommon practice for many women.† Their immune systems were complete opposites; perhaps their instant attraction was an olfactory match made by Mother Nature. After a three-week-whirl wind courtship, Anna and Jacob were married in Lottie's living room by a Congregational Minister. Tecla (a.k.a. Anna), strongly opposed her son's marriage to a non-Jewish woman and refused to attend the ceremony.

Jacob and Anna had three sons in quick succession: Robert, (likely conceived on their wedding night, if not before), Edmund, and my father, Francis Valentine. Anna was pregnant with their fourth child, when she persuaded Jacob to move to Ogden. She was desperate for greater distance from her hostile mother-in-law, Tecla, who was providing her with strong daily doses of critical feedback. Anna was longing for the loving support of her Ogden guardians. After arriving in Ogden, the Liebermann family of five was welcomed into the

---

* Interview with Alice Liebermann Marshall, Salt Lake City, October 1993.

† The New Jersey birth record for Anna Tuite shows her date of birth was October 16, 1886. Her Utah death certificate incorrectly shows October 16, 1888, the date she had long used. Jacob was born in New York City on August 14, 1884.

home of Aunt Lizzie and Uncle Ed, where Anna gave birth to Alice a few months later.

Anna Gertrude Tuite, Ogden, Utah, 1914

Jacob Lieberman, standing behind counter, and Torty Tuite, standing far right corner inside the T & L Coffee House, 137 25th Street, Ogden, Utah, 1918

Jacob went to work for his father's cousin, Simon Bamberger. Unfortunately, Jacob lost the job when he flirted with his supervisor's wife. After getting and losing another job with the Southern Pacific Railroad Company, Jacob persuaded Anna's brother Torty to partner with him in opening the "T & L Coffee House" (T & L stood for Tuite and Lieberman). As the picture of Jacob and Torty inside the T & L Coffee House reveals, Jacob was a good looking man.

Sadly, Jacob exhibited archetypal symptoms of bipolar disease. His seduction of Anna took only three weeks. His hyper-sexuality and promiscuous behavior cost him his job on the Bamberger Railroad and his hot temper repeatedly got him in trouble. According to newspaper reports, Jacob shot and injured a patron in a dispute over a sandwich order. A greater clue, one of more concern, was that Jacob attempted suicide in the T & L Coffee House the day before Valentine's Day 1918.

RESTAURATEUR FAILS IN SUICIDE ATTEMPT

[Herald-Republican Special.]

Ogden, Feb. 13.—Jacob Lieberman, proprietor of a local coffee house, was unsuccessful in an attempted suicide here early today. He was unconscious from the effects of gas when found. Tonight Lieberman is much improved and his recovery is expected. Ill health is assigned as the cause of the attempt.

About 11 p. m. yesterday Lieberman, who lives at 2957 Lincoln avenue, telephoned his wife that he would be home shortly. When at 2 a. m. he had not appeared at the house, Mrs. Lieberman went to his place of business and found it locked. She summoned the police and the door was broken down. Lieberman was found lying on the floor unconscious with a tube running from a gas jet to his mouth.

Salt Lake Herald, February 14, 1918

A few months later, when Anna learned of Jacob's repetitive infidelity, there was a dish-breaking altercation. Instead of making amends and without saying goodbye, Jacob lied to Army officials about his marital status. He impulsively signed up as a single male to serve in the Army of Occupation and was quickly shipped to France.

Neither Jacob's wife, Anna, nor his business partner, her brother Torty, knew where Jacob had gone. Once in France, in a moment of nostalgia, Jacob gave his wife's name and address to a grateful Frenchman, Francois du Verne. He wrote Anna about their meeting and upon receipt of du Verne's letter, Anna ran to the Army recruitment office to notify officers of Jacob's true marital status.[62] Jacob was soon sent back to Ogden.

Jacob's train pulled into Ogden's Union Station on September 20, 1919. Without attempting to reunite with his wife or children, Jacob snuck up behind Hockey Moulding, a young man who had been helping to feed and care for the family Jacob had abandoned, stabbing him in the back. As Anna ran to help Hockey, Jacob ran toward Union Station and disappeared into the night.

In 1921, Anna gave up hope of any reconciliation and filed for divorce in the Second District Court of Utah. The divorce was granted on the grounds of abandonment and neglect on November 15, 1921. Jacob Lieberman never paid the court-ordered seventy-five dollars a month in child support for his four children.

I subsequently learned Jacob's promiscuity did not end when he left Ogden. From old newspapers, I discovered Jacob went to Wappingers Falls, New York, then the home of his younger brother, George Liebermann. In 1999, I interviewed George's twin son, an IBM engineer in Lexington, Kentucky on the telephone.

He said, "I'm sorry to tell you this, but Jake was always messing around with someone else's wife. He even hit on my mother's sister-in-law."[63]

Jacob Liebermann was nothing if not a serial heartbreaker. He was married a second time to Emily Mueller, a woman thirteen years younger, on March 10, 1929. He returned to the original spelling of his surname, Liebermann. The couple had one child, their daughter Jayne.

Jacob was sixty-one when he died in July 1947 of a heart attack outside his gas station on Albany Post Road near Wappingers Falls. Emily lived to be one hundred seven, she was widow for fifty-seven years. Both are buried in the Wappingers Falls City Cemetery.

Jacob Liebermann in 1944 with his second wife, Emily, (behind) and their daughter, Jayne (below), with Emily's sister and mother.

Jacob Liebermann's Gas Station and Store on Albany Post Road, Wappingers Falls, New York, about 1945

As a non-Mormon divorcee in Utah, Anna Tuite Lieberman became a force of nature. She raised and educated four children on her own, under perpetually challenging circumstances. There was a period of several months, when she and Alice shared the same twin bed so Anna could rent out the other twin bed to save money for the education of her children.

Anna's first born son Robert became a successful commercial artist in Chicago. He married and, after he and his wife lost their first born child, they adopted a daughter who was born deaf.

Robert J. Liebermann (Leigh), 1942

Edmund Lieberman with his daughter and son, Chicago 1946

Anna's and Jacob's second son, Edmund, won the Tribune Prize as the top high school student in Salt Lake City. He received an appointment to West Point, but failed the vision test. After a year on a scholarship at Reed College, he subsequently won a full scholarship to attend Harvard where he graduated with honors in business administration.

My father began working at the USDA entomology lab in Salt Lake while he was in high school. Starved for paternal guidance, Frank fell under the spell of the scientific ethos of the men working there and decided to become an entomologist. He applied to colleges his boss identified as having good entomology departments and was admitted to Cornell, Auburn, and Iowa State. As it later did for me, money made his decision. Frank went to Iowa State, the least expensive option, taking his divorced mother and sister with him, working full time to support them while he earned his doctorate in entomology. Alice graduated with a degree in home economics.

As a child I knew Anna Tuite Liebermann as "Grandma Nancy." I remember her ushering my father and me into her private house mother quarters at the Sigma Chi Fraternity on the campus of Utah State University. She was already dressed to preside over the Sweetheart Ball. Her bearing was regal—a perfect blend of dignity and indignity, a mixture that kept everyone around her on their toes. Her gown glittered with gold sequins strung out over layers of brown silk chiffon. When she bent over to offer me a butter mint from the silver bowl on her desk, she smelled like violets.

From my perspective, Grandma Nancy appeared to be living the life of royalty, but she was just a stand-in. In charge of a house full of raucous college boys, her main duty was to maintain a semblance of order within the Sigma Chi house, plan menus, and keep the books. She had to preside at the head table during dinner, but was allowed to take both her breakfast and lunch alone in her private quarters. It was the best job she ever had.

Anna Tuite Lieberman was seventy when she died of a cerebral hemorrhage in her Salt Lake apartment on April 21, 1956. Aunt Alice's son, Bill Marshall, then ten, was staying with Anna the night she died.

Minutes before her death, Anna called out to her grandson, "Bill, the music is beautiful, but could you please turn the sound down."

The only music playing was in Anna Tuite Lieberman's head.

William (Bill) Arthur Marshall and his Maternal Grandmother, Anna Tuite Lieberman, Salt Lake City, 1945

-48-

Converted to the Science of Epigenetics

Since she was the great granddaughter of an entomologist, perhaps it was no coincidence that Tate became deeply interested in Monarch butterflies. Tate, who had just celebrated her fifth birthday, a developmental moment defined by perpetual questions, repeatedly sought answers I did not have. But Tate led me to a new frame for my life because she wanted to understand the mystery of how Monarch butterflies migrate from Mexico to Canada and back—a trip that stretches over several generations.

Tate being visited by friendly butterfly on Nana Joan's back patio, Boulder, July 2003

"Nana Joan, how do the Monarchs know where to go if their grandparents who left Mexico never told them?" she pestered. "Do they leave them some kind of map?"

"I think it must be a cellular memory; somehow they are born with the memory of where to go," I recall telling her.

"But their brains look so small!" Tate responded.

Suddenly my own brain felt smaller.

After Tate's questions, I began to read about how genes are affected by experience and environment, a field known as epigenetics. Scientists have been studying how our health and behaviors are partially determined, not only before our birth, but even before our parents were born. New epigenetic research continues to provide scientific evidence our lives may have been directly influenced by the experiences of our parents, grandparents, and great grandparents—the food they ate or didn't eat, the loving, luxuries, stresses and traumas each experienced.

It seems our genes have chemical attachments that act on our DNA molecules, regulating when, where, and how proteins are made. These chemical additives are markers that change as each of us adapts to our environment, whether in the womb or after birth. They make some of us more immune to stress or disease and others more susceptible to diabetes or biological mental illness. Thus it is possible for these markers to profoundly affect our behavior and also to be passed down to our progeny. Perhaps even more importantly, these inheritances may be modified, even late in life.

Epigenetics turned my beliefs about the role of genes upside down. I no longer saw them as being executives in charge of our development, but rather as raw material controlled at a cellular level by epigenetic processes that are only now beginning to be understood.[64]

In one epigenetic experiment, adolescent mice were subjected to an electric shock at the same time they were exposed to the scent of cherry blossoms. Their offspring also displayed distress when exposed to the scent of cherry blossoms, even though they had not been conceived at the time their parents were subjected to the shocks.[65] The offspring all had larger glomerulus in the olfactory unit of the brain. Reading about this experiment, I thought of the times one or more aspects of my traumatic experiences at Topaz triggered strong physical and psychological responses. And how my daughter and granddaughters are all highly sensitive to smell.

D.W. Winnicott, whose ideas about mothering started me on this narrative, would have been pleased to learn that scientists have now proved that affectionate mothering in rats alters the expression of genes in their offspring in ways that enhance their ability to manage stress. Similarly, epigenetic markers can hinder normal development. The genes of ancestors who experience famine, war, illness, and social stress can silence the gene in their offspring that makes the receptor for the hormone oxytocin, which is the critically important lubricant for the social circuitry of our brains. [66]

During extreme events like famine, migration, and wars, certain genes receive unique epigenetic 'tags' through a process called methylation, a gene modification that deactivates a gene, but does not alter the genetic code. While methylation is part of normal development, patterns vary across individuals.[67] Further, epigeneticists have found that when we are in utero, there is a biological competition between our maternal and paternal genes. Development differs depending on which ones prevail.

Much of my genetic inheritance seems to have come from my father, not only because I didn't develop schizophrenia, but because of my birth as a twin and the genetic mutation linked to breast cancer. On the upside, I inherited my father's scientific skills; unlike him, however, I will not be able to recite the Periodic Table of Elements the week I die. My father kept decay at bay, which may have helped me survive. In two different photographs of my father taken twenty years apart, my father looks exactly the same and his hair was only partially grey when he died at age eighty-six. Whether this was because of his dietary habits and highly restricted caloric intake or something in his epigenetic inheritances remains unknown. Clearly, the risks for depression for my father and me were considerable—risks I have passed on to my progeny.

The development of schizophrenia has been linked to viral infections in utero and to stress in the mother and the child.* I can imagine how terrified Nanie

---

*The effect of an intrauterine viral insult may result in ongoing postnatal damage because of persistence of the virus in neonatal organs. The adverse fetal effects of viral infection result from four primary mechanisms: cell death, abnormalities of cellular growth, chromosomal injury, and secondary

must have been during the pandemic while nursing her sister and her niece. Not to mention the grief and fear she must have felt when ten-year-old Beulah died and she herself fell ill. Recent research has shown how viruses infect neurons[68] and the birth defects caused by the Zika virus may provide further clues about the nature of schizophrenia since scientists are already discovering how the Zika virus crosses the placenta barrier.[69] The complications of my birth and that of my stillborn twin brother, which led to my mother's early oophorectomy and hysterectomy at age twenty-three, were the likely triggers for the onset of her schizophrenia.

In 2014, I found myself wondering whether Dr. Hans C. Moolenburgh was still living; a quick internet search indicated he was. Since I first met Dr. Moolenburgh in the office of Dr. Gonzalez in 1992, he had published a third book entitled, *As Chance Would Have It—A Study in Coincidences.*[*] In the eighth chapter, "Numbers, Dates, Names and Coincidences" Moolenburgh describes how clusters of important events often occur on the same date in families without a particular scientific explanation.

After reading Dr. Moolenburgh's book, I created a calendar of life events for my ancestors, myself, and others who have played a critical role in my life. I also began documenting more carefully the genealogical patterns in the branches of my family tree. As I assembled my ancestral calendar, I was astounded to discover how many events over several generations took place near the same date. For example, my father died April 23, 1998, the same day his mother was buried following her death on April 21, 1956. My mother died on my birthday; Jane Help died on Katherine's birthday. But it was the number of coincidental dates in March that perplexed me.

Fallentin Leibermann married Tecla on March 1, 1874. My mother's oldest sister Vermilla was born on March 1, 1911, six months after Pepa and Nanie married. Army and Nurse were shot by my father on March 1, 1945. While

---

inflammatory responses. Both retroviruses and DNA viruses may disrupt the normal regulation and expression of cellular oncogenes and suppressor genes in the fetus. The viral insult is only one of a series of intrauterine events that lead, after a latent period, to disease.

[*] Moolenburgh, Hans C. *As Chance Would Have It – A Study in Coincidences.* (U.K.: C.W. Daniel Co., 1998), 84-86.

referring to my Berkeley diary, I noticed my appointment with Dr. Maurice Fox, the one that began with talk of suicide and ended with sex, took place on March 1, 1963. The dream-like encounter with the Being of Light occurred on March 1, 1991. A year later, on the same day, Boulder was hit with the "snowy hurricane" that littered the city with broken branches, piercing my unconscious memories of the day the bear went to Topaz. A few days later, the dream of my father filling the dumpster with broken petrified branches led me to take the risk of asking him to reveal more of the secrets of our shared history. It was his willingness to do so that led me to the critical testimony of two other witnesses. I do not know what such coincidences mean. They could be evidence of epigenetic phenomena; Buddhists might see them as indicators of reincarnation; and still others as a validation of their beliefs in astrology.

I can't imagine a scientific procedure that would prove we carry an epigenetic calendar in our cells, but my own history makes me wonder, both about myself and Monarch butterflies, not to mention the still mysterious spawning behavior of salmon who live in the ocean, but return to the remote streams in which they were born to lay their eggs and die. Or the numerous species of birds who follow the same migratory pattern as their parents. There are hundreds of animal, bird, fish and insect examples of what I now see as epigenetic behaviors.

After many years of rumination and research, the only way I can frame my miraculous remission in April 1992 is that some kind of commotion must have occurred at a cellular level, a commotion that activated my battered immune system. I believe that whatever happened was linked to my conscious mind finally being able to face the terror I experienced on The Day the Bear went to Topaz.

Because my life has been filled with many mysteries and fortunate coincidences, my death wish is for scientists to discover more about the cellular and chemical phenomena that create those coincidences, store unconscious memories, and deliver dreams filled with guidance. I will die with both an abiding respect for the wisdom of our unconscious minds and filled with hope for the future of epigenetics.

My unique methylation pattern is the legacy I have already passed on to my progeny. It has taken root in them and will live long after my death. I wish I had taken better care of it, but I did my best under complex circumstances. My highest hope is that the excavation of my life will help them both understand and transform the most damaging patterns of their epigenetic inheritances.

-49-

The Long Worm of My Life

The long worm of my life is still moving between perplexing coincidence and meaninglessness, while trying to avoid the furrows of naivety and paranoia, in my search for meaning.

My mother's preference for ice blue was buried with her; a shade that cannot be found in my home, not even on a hankie. I doubt that the FBI has missed her, having found more dangerous targets.

The surplus Army blankets and Quonset huts of World War II are gone, but brown paper towels are still being used to wipe away evidence of distress, while a single platform rocker, teddy bears, and blankets have been transformed into multi-generational loveys. Although my mother discarded Teda in 1950, I find solace knowing that Orson will outlive me. Thirty-four years after coming to life in our family, Orson sits in his wicker chair, patiently waiting to be loved again, hopefully by a child of Eben Valentine.

Orson patiently waiting to be loved again.

My mother converted the mandated Mormon Temple garments of her childhood into an orthodoxy of girdles, believing it was necessary to wear one even after death. After Mildred Verhaag showed me her garter belt in 1959, I spent a half century trying to muster courage enough for thong underwear. Maybe my genome is on a long trek back to the beginning of time where my descendants will wear nothing, as if born in the Garden of Eden, which Joseph Smith prophesized was in Jackson County, Missouri.

Substitute Mothers: As an old woman, I have gained enough distance to recognize the positive effects of my mother's mental illness. The necessity of being alert produced diagnostic skills that led me to a successful management consulting career. My reaction formation to her fearful paranoia made me generally optimistic and humor has been my best defense.

The psychological damage she inflicted due to her illness was mitigated by a remarkable group of substitute mothers. The first was my maternal grandmother, Nanie, whose love of laughter I swallowed until her humor became a part of my essence. At Nanie's insistence, my father hired June Davis, the young Mormon nurse trainee. Although I do not remember June, I am positive she saved my life, probably more than once.

I still carry the tender nurse-mothering of Bea Wallway like a talisman—the rickrack on her Delta apron became an unconscious lifeline stretching across decades on my self-selected clothing. Bea was one hundred-two years old when she died in her sleep on August 17, 2015, Olivia's birthday.

Afton Evans blessed me, not only with peanut butter sandwiches and Welch's grape juice, but more importantly with predictable tolerant care.

Aunt Mary remained a beacon of love, acceptance, and forgiveness all my life.

Elaine Young's wise anthropomorphism converted me to her fundamentalist feline beliefs. Whenever I open a can of tuna, I still hear her determined voice reminding me, "*Blackie was a heroine, honey, remember that, okay? She died trying to save her babies.*"

Mildred Verhaag tried to teach me the importance of feminine grace and presentation, and, in doing so, reminded me of the mothering I was missing, causing me to break the code of silence about my mother's mind.

In Ouagadougou, Dr. Renee Rebout saved my life by sending me home after we desperately tried, but failed, to save the life of an African mother and her unborn child. Her fearlessness in that emergency helped me save my own life when I awoke to my mother's murderous knife in Tucson.

In Berkeley, I leaned on Professor Calculus and her daughter Rebecca.

In 1971, when I began working at NARF with David H. Getches, I also entered the arena of his wife, Ann Marks Getches. Ann is a remarkable mother and hostess, and I silently adopted many of her rules. On parental curfews for teenagers: "Nothing good ever happens after midnight." For dinner gatherings: "Only one conversation at the table." At Ann's table, I always rise from it filled with awe for her grace, wise ways, and soul nourishing cooking.

When I started motherhood over again, twenty years after Olivia's birth, Martha Penny Hoover, a highly experienced widow-mother, became an essential source of support. Martha's equilibrium and strength in the face of her own losses, helped me stand up when cancer knocked me down.

Martha Hoover showing Eben Valentine why humans rake leaves, October 1985

Martha and her husband, Ted Hoover, were living in Kansas with their three young children when the 1952 polio epidemic took Ted's life and crippled their youngest daughter. Martha moved to Boulder to take shelter with her widowed mother on Marine Street and went to work as a nurse to support her three children. Martha kept her hands on the wheel of life, no matter how rough or steep the road, and she helped me to do the same. Martha was seventy-five when she died while sitting in her favorite chair on February 28, 1998, a few hours before the calendar turned to March 1. When my father died two months later, I felt as if I had lost two beloved parents.

<u>Paternal Care</u>: My father was restrained in his relationship with me, having survived his childhood by avoiding conflict and perfecting his secret-keeping skills. By the time Eben was born, my father started sharing secrets he had been carrying for decades. Frank V. Lieberman's steady servant-like devotion to my children and step-children were acts of love that reversed both of our epigenetic inheritances.

When necessary, I paid by the hour for emergency paternal care, starting with the comforting inspiration of Norris H. Weinberg, followed by the kindness and therapeutic insights of Kenwood Francis Bartelme, James Marquardt, and Roland Evans.

I suspect something about my longing for the open expressions of paternal love may have contributed to the inappropriate distance kept by Professor Hsu and Father Gannon, although I feel ashamed to say so.

<u>My Children</u>: My daughter Olivia introduced me to true love. Her joyful spirit saved and inspired me over and over. Yet, my own scars meant that Olivia was as much a mother to me as I was to her.

As part of her doctoral degree program in clinical psychology, Olivia was required to undergo psychoanalysis. At the time of my expected death in 1991, Olivia asked me to meet with her psychoanalyst when I was in New York.

While Bob and Eben waited on the lawn of the analyst's Long Island office, he spoke to me for Olivia, going directly to the heart of the matter: "Your daughter needed you to be less of a friend and more of a mother!"

Defenseless, the only response I could muster was, "I know."

That night conversing with D. W. Winnicott in a dream, I told him how deeply I regretted my low grade. He kindly reminded me that I had been more mother than daughter to my own mother.

Overcoming this pattern Olivia became an A+ mother to two beautiful daughters.

Olivia's Daughters, Tate and Eve, 2002

Eben Valentine has become as intelligent and helpful as his BaBa. About a year after my father's death, Eben announced his teenage intention to get a tattoo.

"Mom." he said, "I want to have a replica of BaBa's gravestone tattooed on my back so I will never forget him!"

I relented to my son's request, presented with stunning teenage cleverness, despite both Mormon and Jewish prohibitions against tattoos. Not only do I not believe in tattoo taboos, my own body is still marked with one. Now in its sixty-sixth year, my 1951 Atomic Tattoo has somehow miraculously survived the slashing, burning, and poisoning used in the long War on Cancer.

Eben has been as essential to my survival as Olivia, but he may never know just how important he has been unless he reads this book. Because death is a normative topic in our family, he recently promised to do so, adding one qualifier.

"But I don't want to read it until after you are gone, Mom."

Friendships

As an only child, friends have been disproportionately important to me. My first three, Mary Kay Wallway, Marlene Evans, and Linda Olsen, were human floatation devices—they kept me from drowning before I knew how to swim.

Bart Penny was as much my spiritual twin brother as he was my first real male friend. Some part of me still misses the easy coziness of his presence. He fit into my life at critical moments like the missing piece of a puzzle.

Nonny's life has been hard and heartbreaking. In the late 1980's Nonny also battled breast cancer, which was treated with a mastectomy, chemotherapy, and radiation in Eureka. In 1990, Nonny returned to Bakersfield to provide care for her mother until her mother's death on June 15, 1994, the anniversary of Nonny's marriage to Jim. The following year, Nonny created a new home for herself near Taos, New Mexico.

Nonny's greatest loss came in May 2004 when her son Matteo disappeared from his rented residence in near Shelter Cove in Humboldt County, California. Nonny, her daughter, and many friends launched an intensive search for Matteo without success. Two years later, hikers accidentally discovered his jawbone in a remote wilderness area of Humboldt County. An identification was made from Matteo's Coast Guard dental records, but no cause of death could be determined.

Today Nonny continues to reside near Taos, where she provides psychic readings, as well as helping and being helped by friends. She has remained free of cancer.

Joan Carol and Nonny in the dream house garden, Boulder, October 2016

No close friendship is risk-free. We risk rejection whenever we move close to another person. At age seventy-five, Nonny and I carry old relational wounds, but our friendship and emotional attachment endures.

My beloved friend Charlotte was five-years old when her parents, Rubin Sternberg and Frances Laura Segal, emmigrated with her and their son Emanuel from Lviv (a.k.a. Lwow or Lemberg) to America. It was 1939, a very fortunate year for a Jewish family to leave Poland.*

Rubin Sternberg and Laura Francesca Segal and their two children, Emanuel and Charlotte, Lviv, Poland, January 1939

After college, Charlotte married Howard Smokler in New York. When he joined the CU philosophy faculty in 1970, they purchased a home on Marine Street across from mine. She and I have now been close friends for fifty years. Charlotte is still the most intuitive and insightful person in my life. One

* Lvov is now part of Ukraine.

example, during the impeachment trial of President Bill Clinton Charlotte said: *"Five hundred years from now human beings will know all about God, understand time, and events like the Big Bang, but sex will still be a complete mystery."*

Before her retirement, Charlotte worked for many years as the librarian and book reviewer for the *Boulder Daily Camera.* Today she continues to live in her home on Marine Street. Once, when she was in New York after the death of her brother, I found myself trying to be as helpful as my father. I moved too close and crossed several of Charlotte's boundaries without adequate permission. Fortunately, she quickly forgave me. I am grateful, not only for Charlotte's insights and forgiveness, but because she saved me from some of my worst tendencies by marking up the first draft of this book.

The ONINers became the large family that I had often yearned for as a child. When we were no longer working together at NARF, we began gathering to celebrate holidays and birthdays, particularly when one or more of us crossed a forty or fifty mile marker. Those birthday parties were raucous roasts, with skits and costumes, staged in elaborate settings. Like siblings, we were fully familiar with each other's foibles—including quirky habits, painful divorces and remarriages, espoused values incongruent with actual behaviors, and challenging children.

On the fiftieth birthday of David H. Getches, our collective trio of Pulik, (Marley, Dakota, and Dusty), herded the ten of us into the Collegiate Range of Colorado's Middle Park, where we established an elaborate "Camp David." Then, in honor of his mother's dream that David would become President of the United States, we rigorously vetted him for an imaginary upcoming campaign.

For Bob's fiftieth, the ONINers served him a deli breakfast with only the *New York Times* for company. After a lox and cream cheese omelet, they drove him into the country east of Boulder for a session with a psychic, before accompanying him on a clothes shopping expedition. After a massage, Bob returned home for dinner where fifty yahrzeit candles were burning around the dining table and the celebratory menu memorialized every woman in his life.

For my fiftieth birthday, the ONINers gathered at the Broadmoor Hotel in Colorado Springs where they staged a trial to consider whether I should be re-admitted to the Mormon Church.

On my sixtieth birthday, I was publicly interrogated by The Church Lady, a part played to perfection, not by Dana Carvey, but by David H. Getches. I laughed so hard I wet my pants.

David H. Getches as the Church Lady on author's sixtieth birthday, 2002

The ONINers with their children. First row, left to right: Richard Collins and Hilary Reid-Collins, Sasha and Bruce Greene, Audrey Hart, Susan Rosseter Hart, Eben and Joan Carol, Ann Getches, Ben and Charles Wilkinson. Second row, left to right: Oriana Reid-Collins, Judith Reid, Ethan Greene, Jay Hart, Olivia, David Wilkinson, Jordan and Bob Pelcyger, David Getches, Liza Getches, Cybele Reid-Collins, Philip Wilkinson, Ann Amundson, Gale Pelcyger, and Catie Getches. Koenig Center, Boulder, August 1991

Since our work together in the early 1970's, the ONINers have grown grey, traded in some of our original joints for better equipment, and taken delight in the births and achievements of our collective children and grandchildren. Sadly, in 2003, one couple stopped coming to our ONINer gatherings after a son developed schizophrenia; his care took priority. In 2004, another couple unexpectedly separated and divorced. In 2011, David H. Getches died only thirty days after being diagnosed with pancreatic cancer. After those losses, the parties diminished in frequency and increased in simplicity as rich food and wine were no longer well-tolerated—the empty places at the table harder and harder to face. I have been more deeply affected by the loss of these connections than other ONINers, perhaps because they were more like family for me than I was for them.

Twins: Six centuries before I was born, the mythical French twins, *Valentin and Orson,* arrived in a fairytale. When I think of meaningful coincidences, I often think of Gale, who inadvertently introduced me to the story by sending booties in a small gift box illustrating that fairytale. Two decades later, Gale gave birth to in-vitro twins, a boy and a girl.

Further back on the trail of twins, Bridget Monaghan Tuite gave birth to twins. Her twin daughter Anna Gertrude survived, but her twin son, Amico, died on his thirteenth day of life. Twelve years later, Bridget died while pregnant with a second set of twins.

Jacob Liebermann, barely survived his own birth; his twin brother did not, nor did my own twin brother.

I must not forget the premature twins, a boy and a girl, that Bea Wallway birthed and lost in Delta, only a few days after she brought me back to life after the Bear Went to Topaz.

The first Colorado home of David H. and Ann Marks Getches was located on Twin Sisters Road, where in 1976 Ann became pregnant with twin sisters. My mother crafted twin baby quilts for Ann—highly unusual and perplexing behavior on her part. Sixteen years later I finally learned my mother's interest in twin births was not random. David and Ann became grandparents of twin grandsons in 2005.

Ten days after David's death, Katherine Help gave birth to twins, a boy and a girl, whom Bob and I love as our own grandchildren, trying to imagine what they would have meant to Jane.

Canines and Felines: Laid out before me are precious shards of cross species experiences with canines and felines. Contentment comes with knowing my children and grandchildren share my deep love and appreciation of animals.

My canine relationships began with the Delta-born springer spaniels, Army and Nurse, innocently executed on March 1, 1945 because they tried so hard to protect me. They were resurrected in a dream and came back to remind me of what I needed to know to heal. My life will end with the loving comfort Teddy, who carries half their genome in his spaniel-like ears. In between these canine book ends, Pablo demonstrated cross-species tolerance, while magical Moxie was religiously devoted to Eben. Never to be forgotten is Marley, the Puli. Somehow her mother-like mind read my own.

As for felines, beginning in 1948 Snowball modeled what it meant to be a mother under less than optimal conditions, while Blackie's behavior reminded me that I was not the only abandoned creature in our world.

Olivia's beloved Grey Cat demonstrated cross species love and trust with kisses for Pablo and Moxie, while wiping our human tears with her tender tongue. A mother bear with twin cubs made a meal of Olivia's Tiger Cat before Manx took his place and became a star in the hall of feline neurotic behavior with his own psychoanalyst. Each cat taught me invaluable lessons using radically different methods. I miss them all still.

Insects: There is a swarm of insect shards from beginning to end of my worm-like life. Before my birth, Mormon crickets devoured the first critical crops of my maternal ancestors in the Salt Lake Valley. My father, drafted to command a War on Weevils, armed with DDT, also had to fight cockroaches in the Southern Hotel before being deployed to fight the lice, bed bugs, and desert scorpions at the Topaz Relocation Center

After our move to Logan, my father undertook a special one-time assignment to eliminate an onslaught of gypsy moth caterpillars on behalf of his traumatized daughter, just before the federal government sent him into the treacherous atomic testing zone of the Cold War to determine how many of America's pollinators had survived.

An alarming swarm of locusts drew us out of my mother's Dream House, away from Mormonism, into Montana. Those locusts were quickly subsumed by the costly and voracious hunger of spotted alfalfa aphids rampaging through California crops.

Although I haven't heard warnings about South American killer bees for decades, the USDA's decision to send my father to Tucson to prevent killer bees from crossing into the United States was ironic because soon his wife was going toward killer bees trying to hide from the FBI in Nogales.

In Ouagadougou, the bite of a single mosquito carrying the yellow fever vector brought me close to death, and changed the course of my life. On Idaho's Nora Creek Road, I was drafted into an all-out assault on an army of occupying spiders which had wrapped an abandoned house in webs. My granddaughter's intense interest in the lives of Monarch butterflies woke me up to the science of epigenetics, helping me understand my own spiritual migrations and giving me faith in her future.

My life is ending with the emerald ash borer, an illegal immigrant from Japan, now threatening ten thousand ash trees in the Boulder Valley, including two large ones sheltering our dream house, plus the forty or more shading the ashes of my father and soon mine as well. I hope the parasitic wasps being imported from China on special visas to battle the borers will prevail. We live in a world without protective boundaries. The distance between disaster and survival is ceaselessly changing.

Couches: The largest couch shard is the Dinwoody's couch purchased by my parents in 1941. My mother's madness stuffed it with poisonous snakes in Delta before it was moved to Logan, where it was amputated. Then Mayflower Moving men transported it first to Bozeman, and a year later to Bakersfield, where it was abandoned still filled with poisonous snakes when my parents fled to Tucson in 1961.

There was the too-short brown mohair couch in the living room of the Evans family that led Ben Evans to ask Afton for his birthday couch. After a single nap, Ben's birthday couch was turned into a plastic-covered monument to their marital stalemate.

I woke up to my mother's murderous intentions on a Tucson daybed, which, along with its twin, soon became the Tucson Torture Racks inside my immorality prison where they served as unconsummated marital beds. After becoming pseudo couches, I sat upon one while I sought a divorce, before abandoning both at a free-furniture site in Oakland.

Lary Carpenter created three unique pseudo-couches, but our marriage ended with a ten dollar twin of the Dinwoody's couch serendipitously discovered at a Volunteers of America store in Denver.

The antique couch I bought without Bob's permission was moved to Oregon by Katherine Help before she brought it back to Boulder along with her twins in 2016.

I stole secret naps on Ruth's white satin brocade couch without staining it. Now, Bob and I snuggle on a Crate and Barrel loveseat upholstered in brown mohair. Just like the brown mohair couch in the living room of Ben Evans, it is too short for a comfortable nap.

Birds: When I close my eyes I can still see the watchful owl eyes of Olly glowing in the dark as Bart and I lay on a Bozeman bed. Thirty-five years passed before I dreamt in my Boulder bed of birds in Bozeman. Magically, the very next day I was reintroduced to the birds of the Great Salt Lake by Terry Tempest Williams in *Refuge.*

Terry's eloquent narrative was a reverent return ticket to my Mormon past and the summer days I spent near the Bear River Refuge, where, perched bird-like, I competed with seagulls to harvest cherries from trees so I could run away from home. *Optimal Distance* was laying abandoned in the nest of my mind until *Refuge* broke open the shell around my psyche.*

A quarter of a century after my metaphysical experience along the shore of Yellowstone Lake, I rarely go a day without thinking of the trumpeter swan, who like a feathered angel swooped down from the sky to trade her life for mine.

From inside our dream house, I savor watching the magpies and crows who have learned to recognize my husband, as *Corvidae* cousins are want to do. With their raucous chorus, they celebrate my husband's appearance in our rear garden because he lovingly brings them banana chips carefully picked out of his morning granola.

Wars: The shards of many wars are strung along the long worm of my life, beginning with the military deployments of World War I, which spread the Spanish Influenza, turning it into a pandemic that infected Nanie, before crossing her placenta barrier to invade my mother's mind while she was still in utero. In Ogden, only forty miles away, Jacob Liebermann used the World War I Army of Occupation in order retreat from a marital conflict in which Anna

---

* Nine years later, I was stunned when I read *Leap,* (New York: Pantheon, 2000), Terry's deep exploration of Hieronymus Bosch's painting, *The Garden of Earthly Delights*, Bob's favorite which has graced our walls since 1972.

Tuite used her only set of dishes as weapons of retaliation against his adulterous betrayals.

My father was saved from death or maiming in World War II by being drafted into the War on Weevils, dangerously armed with lethal DDT. Had I been born in Europe instead of America, I would have been declared a *mischling* and tossed into one of Hitler's ovens.

I learned about the true nature of war between human beings in The Great Fried Egg War of 1949.

Twice lucky, as a Gentile, I wasn't given my Atomic Tattoo until the Cold War.

My pseudo-marital vows kept Lary Carpenter from being killed in the War in Vietnam before I signed up to serve in the War on Poverty.

I was already fodder in the War on Cancer, when the Gulf War inspired me to write a vision for Friends' School.

My husband is still fighting in the War on Native Americans, where modern weaponry includes bureaucratic blockages, fake news, and flagrant violations by the Federal Government of trust responsibilities to tribes.

My life has been extended by artificial arteries still being perfected in the War in Iraq. That conflict, as well as the War in Afghanistan, the Syrian Civil War and the War on Terrorism now occupy the contextual history of my progeny.

When I think about the millions of victims of these conflicts, I am reminded of what I learned in Paris from Claude Levi-Strauss, the French anthropologist. Lecturing at the *Université Paris-Sorbonne*, he told the story of two primitive tribes living on opposite sides of a wide river in a far off place. The members of the tribe on the south side of the river called themselves "The Great One's Special People" and referred to the tribe living on the opposite bank as "God's Shameful Mistakes."

Crossing the river to visit "God's Shameful Mistakes," Levi-Strauss learned the tribe referred to themselves as "The Greatest Ones" while they referred to the tribe on the south bank of the river as "What God Spit Out."

A half-century later, as I watched American voters elect President Donald J. Trump, I thought again of Claude Levi-Strauss. Our country has two tribes living on opposite sides of an ideological river that seems to be growing wider, flowing faster. We are in desperate need of strong swimmers with courage enough to attempt crossing that wide divide.

Bears: Twin bear cubs have repeatedly crossed my path. Beginning with the mother bear and twin cubs who so wanted my mother's cherry chocolates; to the mother bear with twin cubs who raided our borrowed house on Nora Creek Road; to the mother bear with twin cubs who pursued me along the shoreline of Yellowstone Lake.

Since I was eight I have had many powerful dreams about bears. Now hungry bears visit my own dream house, the interior of which is heavily populated with ursine totems, including Orson.

Ursine Twins Rising Early in Yellowstone, 1957

The Mother Bear of Twin Cubs Begging on Fish-Car, Yellowstone, 1957

Death: If we are lucky, we first meet the finality of death in an animal or a bird. It started for me with Blackie's first litter, followed by six male kittens instinctively slaughtered by a tomcat. But I didn't really feel the finality of death until I wrapped Blackie's stiff corpse in the blanket of Elaine's baby boy. I moved through frogs and felines in zoology before helping Dr. Rebout wash and wrap the still warm corpse of an African mother and her stillborn son. The latter was preparation for repairing the make-up of my own embalmed mother. After the deaths of Jane Help and Ruth Pelcyger, I dressed their bodies without fear. I stroked the hair, hands, and feet of my father's corpse, before I carried away a lock of his hair, erroneously thinking I could let him go. More recently, I staggered with grief as I walked alongside a gurney carrying the body of David H. Getches from his kitchen death-bed to a waiting hearse for his final ride to the crematorium—a ride my remains will soon take as well.

Spiritual Legacy: Yellowstone became my cathedral of worship, providing awe and mystery. Whenever I revisit its vistas in my mind's eye, I feel restored.

I feel no worry about having to stand before an all-powerful Heavenly Father as he opens his notebook to count the black marks on my page. In this respect, I am an atheist for whom no such deity exists. If I am wrong, and end up in some Hell, I will owe both the Prophet Joseph Smith and Bishop Marc Ricks an apology.

While I have not achieved *optimal distance* from everyone in my life, including some who have died before me, I am no longer anxious about the close proximity of death or distressed by my mixed moments of atheism and agnosticism. Though I am surrounded by Buddhists in Boulder, I do not believe I will be reincarnated in any form.

As a recent convert to the science of epigenetics, I am strangely comforted by the idea that I have already left my epigenetic legacy to my children and grandchildren. Despite everything, including whatever cellular methylation they inherited from me, they are doing well in the world. I hope my history will help them and other readers to understand the epigenetic patterns they have inherited and are likely to transfer to future generations.

Surrounded by shards from the historical heap of my life, I have set aside my shovel and sieve. The proximity of death does not detract from the deep sadness I experience whenever I look into the eyes of my husband or snuggle against his warm body, or have long intimate telephone talks with Olivia, or watch Eben playing with Teddy, or my tearful feelings whenever I think of the grace and intelligence of my granddaughters. I do not want to say goodbye, but I believe it is a gift to die with the kind of yearning I feel for them and others.

-50-

A Final Dream

In the fall of 2014, Boulder experienced an unusually long and beautiful Indian summer. Trees were holding their colorful leaves; temperatures were mild enough that ten days into November fragile begonias were still blooming in our mountain-side garden. But on November tenth a polar plunge began at noon and by midnight the temperature had dropped seventy-seven degrees, the third largest one-day temperature fluctuation on record for the Boulder Valley. After temperatures fell to minus ten, snow began to fall. Temperatures remained below freezing for several days. The trees and gardens in Boulder resembled those of Pompeii, frozen in place, not by hot ash, but by ice. More than ten thousand trees were killed by the shock. Along the north side of our dream house, a twenty-year-old yew hedge as tall as Ben Evans died overnight.

Big Boy Bear on back patio of dream house 2015

The polar plunge was followed six months later by an unusually late freeze on Mother's Day 2015, the conceptual anniversary of this book. The freeze killed most of the blossoms on fruit trees including apples, plums, and pears, so there was no summer fruit on the boundaries of Boulder. By late summer, the bears living in the mountains west of the city were starving. Our motion-activated-camera captured a large male bear precariously balanced as he stood on the narrow wrought iron railing of our garden bridge, desperately searching for a single wild plum on the branches just beyond his reach.

I began calling him Big Boy. My heart aching, I emptied our freezer of frozen berries and salmon filets, illegally serving them up on the far side of our flimsy fence. Big Boy came every day, often before dark, searching, then sitting quietly, his expression one of gentle expectance. He was unusually polite and well-mannered. My last sighting of him that year was in November, just before Thanksgiving.

Since then I have had a repetitive dream in which I find myself sitting down next to him, while he cuddles a swan in his arms beneath the Angel Tree.

Epilogue: For My Husband

My marriage to Bob Pelcyger has now lasted forty-two years. Our marital boat has sailed through some dangerous storms. Despite lacking a compass and being tossed and turned, neither of us has fallen or jumped overboard. Were it not for my husband, I would be dead—my existence a few cellular memories in my progeny. Not only has Bob never abandoned me, his spousal devotion has kept me alive.

When Bob and I first met, I was as short of trust as Manx, the cat. Bob had an immediate effect on me, but, unlike Manx, I did not immediately turn into silly putty. Over several years of courtship Bob slowly won my complete trust. As I dropped my defenses and fears, one after another, Bob came to know all my faults, large and small, as well as my sins, worries, and dreams. He is a remarkably tolerant soul whose patience for solving a problem, no matter how long it takes, is remarkable. Just ask those Native American clients he has represented for fifty years. It is not surprising that Bob has been adopted as member of the Crow Tribe of Indians, as well as by a flock of crows who follow him on the streets of Boulder. Children, animals, and birds somehow sense what Manx showed me—Bob Pelcyger is an extraordinary man.

Our blended brood has been safely delivered to the shores of adulthood. Three have found marital partners and have children of their own. Eben has yet to do so, but he has found passion in his work. While it is unlikely that I will live to see him married, watching Eben's loving care of his nieces, nephews, and Teddy has given me confidence that, if he so choses, Eben will be a wonderful parent. Katherine Help, who seems as if she is our own child, is a gentle mother in the model of Jane, one who is traveling a steep road, carrying the full load of single parenthood while she finishes her doctoral degree. Bob and I watch her strength and endurance with awe, trying to provide the parental support she needs.

Teddy waiting to snuggle.

Although age-eligible for the dreaded assisted living consolidation, Bob has been insistent that we age-in-place in our dream house. Each day after Bob wakes me, I weave, sometimes wobble, my way to the kitchen where he has laid out my morning medications, before carefully negotiating the two steps up into our living room. As I wait for the morning cocktail of life-supporting pharmaceuticals to disperse throughout my body, I make myself comfortable in a large chair where I can look out on the terraced garden and the stone steps leading up the slope of Boulder's Flagstaff Mountain. Teddy jumps up to snuggle beside me; his soulful eyes and the warmth of his body instantly sooth my heart. I savor each moment of these peaceful mornings. As I watch the passing wildlife work to survive, I do the same.

In 2014, Bob read my *Optimal Distance* manuscript and our relationship experienced a late-life renaissance. We began talking about our lives from new perspectives. The distance between us shrank. As our psychological closeness increased, we renewed our physical intimacy; I even purchased new lingerie. Each time a Google advertisement reminds me of those purchases, I think of Ruth and Maxwell.

My personal history would never have been written without the tapestry of love and care Bob has woven around me. Nor would I have had the courage to undertake such a deep excavation of its numerous layers. Bob's generous and tireless support and my deep love for him combined to create *Optimal Distance*.

Acknowledgements

This narrative provides glimpses of the help I received from others—those still living and others now dead. Some gave me a smile, words of encouragement, a cup of warm milk, an infusion of hope, or a tender embrace. Others shared their recollections and insights in writing. Some recommended a book or gave me the name of a connection. A few gave me courage by asking for my help. All were instrumental in my unexpectedly long survival and kept me from being lost.

Bread Loaf Writer's Conference provided a space for Blackie and Snowball to come back to life with the encouragement of Patricia Hampl, as well as Carol Houck Smith of W.W. Norton.

Distant librarians responded to my esoteric inquiries and both family members and my friends tolerated my long research process.

Scott S. Miller kept the wheels from flying off the complex task of preparing a two-part book for publication. He repeatedly provided emergency technical aid in the middle of the night, undertook the restoration of old family photographs, and captured new images.

T. Keith Harley's eye for the visual, combined with his exquisite design talents, brought both Part One and Part Two of the manuscript to life on the covers and his designs for the interior pages.

Kim Brown Federici, Jennifer Breman, and Laura K. Hink were early readers who helped me to persist. The corrections and comments of Charlotte Smokler, Ann Marks Getches, and Roland Evans became gifts of guidance.

Above all, I am filled with gratitude for the remarkable editing talents of Emma Komlos-Hrobsky in New York City. Emma, in addition to being a wonderful writer, is an unusually careful reader. Her therapeutic insights and suggestions brought me closer to the core of my life.

Glossary of Individuals and Animals

* = *animal*

Al-Madani, Sidi Tayyib. Osama Bin Laden follower who betrayed him.

Amundson, Ann. ONINer and wife of former NARF colleague Charles F. Wilkinson, who bought *Refuge* for author; subsequently lent her other books by Terry Tempest Williams and arranged for them to meet at the Pearl Street Inn in 1992.

Andre, Susan. Friend and second wife of Bruce R. Greene.

Angel Moroni. In 1823, Joseph Smith reported the Angel Moroni led him to the Golden Plates, the source material for the *Book of Mormon*.

Army.* Author's first dog, a male springer spaniel.

Ashby, Mrs. Name listed in Bea Wallway's 1992 letter to the author.

Ashby, Roger W. Witness on Day Bear Went to Topaz.

Audeh, Nuhiela. Metastatic breast cancer patient whose radical remission led to informal clinical trial with potion of Thuja or the Tree of Life.

Aunt Mary. Maternal aunt of the author, see Beck, Mary Isadore.

BaBa. Eben Valentine's nickname for his maternal grandfather, Frank V. Lieberman.

Bamberger, Simon. Jewish businessman with railroad and mining interests. In 1917, he was the first non-Mormon Democratic governor of Utah and remains the only Jew elected to that office. First cousin of author's paternal great grandfather.

Bartelme, Kenwood Francis. Clinical psychologist who treated author in Berkeley in 1965 and briefly in 1967.

Beach, Andra. Friend of Jane Help whose son Evan was Adam Help's playmate.

Bear, Mother with Twins.* Yellowstone bear who chased author along lake shore, but settled for swan in October 1992.

Bear, Big Boy.* Male bear who visited author's dream house in fall 2015.

Beck, Jack Campbell. Author's maternal uncle.

Beck, Margaret Audrey. Maiden name of author's mother, a.k.a. Mugs.
Beck, Mary Isadore. Author's maternal aunt, substitute mother.
Beck, Edward Robert. Author's maternal grandfather, a.k.a. Pepa.
Beck, Vermilla Smoot. Author's maternal grandmother, a.k.a. Nanie.
Beck, Vermilla. Author's maternal aunt, born March 1, 1911.
Bellows, Warren E. Acupuncturist who once treated the author.
Berrens, Joy. Equestrian who befriended Addie Help after Jane's death.
bin Laden, Abdullah. First born son of Osama Bin Laden.
bin Laden, Khaled. Son of Osama and fourth wife, Siham Sabar
bin Laden, Osama. Mixed-blood Sunni Arab terrorist who planned September 11 attacks, founder of al-Qaida.
Bindler, Sarah Rebecca. Maternal grandmother of Bob Pelcyger.
Blackie.* Author's female cat in Logan in 1948-1949.
Boehmer, Robert Nathan. Husband of author's first cousin Lynn Marshall Eaton, daughter of Aunt Alice.
Bosch, Hieronymus. 15th century painter of *The Garden of Earthly Delights.*
Briggs, Susan. Author's EBHS classmate, who traveled to Europe with her.
Brown, Kim. Author's EBHS classmate, who traveled to Europe with her, now Kim Federici.
Budge, Dr. Omar Sutton. Physician and "x-ray specialist" who treated author in Logan, Utah 1951-1956.
Butcher, Charles. Owner of Butcher Wax Company, benefactor of the Thuja trial.
Calculus, Rebecca. Daughter of Professor Calculus who helped care for Olivia in Berkeley, 1963-1965.
Calculus, Professor. Professor of Mathematics at UC Berkeley, author's professor, neighbor, and close friend.
Callie. See Tempest, Callie
Cantor, Samuel. Maternal grandfather of Bob Pelcyger, born Minsk.
Carrie. First cousin of Deborah, Bob Pelcyger's first wife.
Christy. Young mother who switched airline seats with author in July 2007 on flight to Salt Lake City.
Clifford, Glenn. Husband of Margaret Martin Clifford, family therapist, and Friends' School parent,
Clifford, Katie. Daughter of Margaret Martin and Glenn Clifford.

Clifford, Margaret Martin. Friends' School parent who became a close friend of the author while both were being treated for metastatic breast cancer.
Cole, Bev. Helped author found Friends' School in 1988, mother of Cole Davis.
Collins, Richard B. ONINer and former NARF colleague.
Cousin Julie Ann. Maternal first cousin of author; daughter of Aunt Mary.
Cushman, Dr. Florida cardiologist who treated Carrie.
Dakota.* Help Puli puppy adopted by Getches family.
Davidson, Max. Jewish owner of cigar factory, who employed author's paternal grandmother in Ogden, Utah.
Davis, June. Student nurse hired to care for author in Delta, Utah in 1942 until June 16, 1944. She likely saved the author's life on more than one occasion.
Day, Dr. John. Surgeon who treated Nuhiela Audeh and Thijs De Haas; helped supervise Thuja trial.
DeHaas, Thijs. Scientist who brought initial supply of Thuja to America.
DeHaan, Trina. Aunt Alice's granddaughter, second cousin of author, friend of Callie Tempest.
Deborah. First wife of Bob Pelcyger; mother of Gale and Jordan.
Denevere.* Female Puli belonging to Help family.
du Verne, Francois. French citizen who met Jacob Liebermann while he was serving in the Army of Occupation and wrote Anna Tuite Lieberman in August 1919 about their meeting.
Duncan.* Male Puli given to Addie Help in May 1989 for her sixth birthday.
Dusty.* Help Puli puppy adopted by Wilkinson family.
Eaton, Craig. Author's second cousin; Aunt Alice's grandson.
Eaton, Lynn Marshall. Author's first cousin, daughter of Aunt Alice.
Eikenbary, Susan. Friend of Jane Help, Boulder, 1992.
Einhorn, Dr. J. USDA scientist who brought first Pulik to America in 1939.
Elaine. Substitute mother and life science tutor in Logan, Utah, 1948-1950. (See Young, Elaine).
Eppinger, Catherine. Marley's veterinarian.
Epstein, Bob. Friends' School parent and artist.
Evans, Afton Lee. Substitute mother to the author in 1947-1948.
Evans, Ben. Father of Marlene Evans; author's iconic model of equanimity.
Evans, Marlene. Author's best friend, daughter of Afton and Ben Evans.

Evans, Roland. Irish-born psychotherapist who provided hypnosis to author in Boulder during her treatment for metastatic cancer 1990-1992. He has continued to provide psychotherapy to author, Bob, and Eben since then.

Eve. Author's second-born granddaughter.

Getches, Ann Marks. ONINer and wife of David H. Getches, close friend.

Getches, Catie. Twin daughter of Ann Marks and David H. Getches.

Getches, David H. First director of NARF, ONINer and husband of Ann Marks, close friend of author and her husband.

Getches, Liza. Twin daughter of Ann Marks and David H. Getches.

Gill, Stanley Jensen. CU Professor of Chemistry, Thuja clinical trial participant who died June 15, 1991.

Gold, Larry. Husband of Hope Morrissett, father of Nicholas Gold, and benefactor of Friends' School. CU Professor of Molecular and Cellular Biology and founder of several biotech firms, who assisted author in arranging for the clinical trial of Thuja in 1989.

Goldberg, Dr. Helen. Author's favorite oncologist who practiced briefly in Boulder in 1993-1994.

Goldschmidt, Tecla Levy. Author's paternal great grandmother.

Goldstein, Arthur, Esq. Brooklyn lawyer who employed Ruth Cantor and Eugene Pelcyger in the late 1930s.

Gonzalez, Dr. Nicholas. Physician who used metabolic therapy to treat cancer and other conditions. Author was his patient from 1990-1994.

Greene, Bruce R. ONINer and former NARF colleague.

Greene, Ethan. Son of Reggie Gray and Bruce R. Greene.

Greene, Sasha. Daughter of Bruce Greene and Susan Andre.

Grey Cat.* Olivia's female calico cat who was born in barn near Troy, Idaho in 1967.

Griffin, Jeffy. First head teacher of Friends' Primary School, 1988-1991.

Grothjan, Naomi. Principal of University Hill Elementary, Thuja clinical trial participant.

Hamidi. Syrian-born mother of Osama bin Laden.

Harris, Edmund Joseph. Husband of Elizabeth Esther Tuite, a.k.a. Uncle Ed, foster father to author's paternal grandmother in Ogden, Utah.

Hart, Audrey. Daughter of Susan Rosseter and Victor Hart.

Hart, Jay. Son of Susan Rosseter and Victor Hart.

Hart, Susan Rosseter. ONINer until 2005; worked for author at Project Head Start, and was third wife of Bruce R. Greene. They divorced in 2005.

Helmke, Paul. Author's client while mayor of Fort Wayne, Indiana.

Help, Adam. Eighth child, second-born son of Jane and Tom Help.

Help, Addie. Seventh child, sixth-born daughter of Jane and Tom Help.

Help, Christina. First-born child and daughter of Jane and Tom Help.

Help Family. Jane and Steven Help were the parents of eight children. Jane developed metastatic breast cancer at same time as author.

Help, Jamie. Second-born child and daughter of Jane and Tom Help.

Help, Katherine. Sixth child, fifth-born daughter of Jane and Tom Help.

Help, Lizzie. Fourth-born child and daughter of Jane and Tom Help.

Help, Maddie. Jane's first-born grandchild, Lizzie's daughter.

Help, Ronald. Fifth child, first-born son, of Jane and Tom Help.

Help, Tara. Third-born daughter of Jane and Tom Help.

Hirschhorn, Larry. Consultant colleague and book editor, Wharton Center.

Hill, Anita. Employee of Clarence Thomas before he was appointed to U.S. Supreme Court in 1991.

Hoover, Martha Penny. Neighbor, substitute mother, guardian of Moxie.

Hoverstock, Father Rol. Rector, Boulder's St. John's Episcopal Church

Hussein, Saddam. Autocratic fifth president of Iraq.

Hyacinth. Jamaican-born nurse who helped care for Ruth Pelcyger.

Ina. Ruth Pelcyger's friend in Hollywood, Florida.

Isaacs, Dr. Linda Lee. Physician, first wife and partner of Dr. Nicholas Gonzalez in New York City. She treated Carrie in 1993.

Jeanne. Widowed mother of Jane Help.

Jennings, Peter. ABC Evening News Anchor

Jennings, Dr. R. Lee. Author's tumor surgeon, 1986-1989.

Jordan, Ginny Wells. Mother, therapist, and writer; introduced author to treatments offered by Dr. Nicholas Gonzalez in Boulder, 1990.

Katz, Jesse. First friend of Eben at Mapleton Elementary, son of Kathy Mackin and Andy Katz.

King, George. Ruth Pelcyger's brother-in-law, husband of Selma Cantor.

King, Selma Cantor. Ruth Pelcyger's sister, wife of George King.

Krantz, James. Author's colleague from Wharton; took over responsibility for Fort Wayne Vision Project in 1991.

Leibermann, Fallentin. Author's paternal great grandfather.

Leigh, Robert Jacob. Author's paternal uncle, who changed his surname from Liebermann to Leigh.

Levi-Strauss, Claude. French anthropologist; author heard him speak in Paris in 1962.

Liebermann, Alice. Author's paternal aunt, also known as Aunt Alice; married Arthur William Marshall in 1941.

Lieberman, Donald. Twin son of George Liebermann, author's paternal great uncle.

Lieberman, Edmund. Author's paternal uncle and Harvard graduate.

Lieberman, Frank V. Author's father, born Francis Valentine Liebermann.

Liebermann, George. Brother of Jacob Liebermann, paternal great uncle.

Liebermann, Nancy. Nickname for Anna Tuite Liebermann.

Luce, Dr. David. Internist who treated author while she was on Gonzalez protocol, 1990-1993.

Magic.* Male cockatiel belonging to author's son in 1993.

Manx.* Son of Grey Cat, who was born with stubbed tail in Boulder 1970.

Marks, Charlotte Anne. Wife of David H. Getches and hostess phenom, also known as Ann M. Getches, ONINer and close friend of author.

Marley.* Female Hungarian Puli, born on birthday of Bob Marley, who was fifth birthday gift to author's son in 1988.

Marquardt, Dr. James E. Author's Boulder psychiatrist, 1973-1983.

Marshall, Alice Liebermann. Author's paternal aunt.

Marshall, Arthur William. Author's paternal uncle by marriage.

Marshall, William Arthur. Author's first paternal cousin, son of Aunt Alice, also known as Bill Marshall.

Martin, Cheryl. Sister of Margaret Martin Clifford.

McKenna, Charlotte. First cousin of Anna Tuite Lieberman, her nickname was Lottie.

Miller, Barbara B. MSW. Family therapist who worked with author's family, Boulder, Colorado, 1990-1993.

Miriam. Widow of rabbi and conference coordinator who assisted author in Salt Lake City after stroke accident and surgery in July 2007.

Monaghan, Bridget Author's paternal great grandmother.

Miller, Annie. Neighbor in Delta, Utah in 1946.

Moolenburgh, Dr. Hans C. Dutch physician and author, who reviewed author's case and examined her after her radical remission in 1992.

Moran, Dr. Pat. Oncologist who attended *Thuja* clinical trial meeting.
Morrissett, Hope. Helped found Friends' School; wife of Larry Gold and mother of Nicholas Gold.
Moulding, Hockey. Friend and admirer of Anna Tuite Lieberman and her four children when they lived in Ogden, Utah. His given name was Harland.
Moxie.* Male golden retriever who became Eben's guardian. He belonged to author's next door neighbor, Martha Penny Hoover.
Mueller, Emily. Second wife of Jacob Liebermann.
Murphy, Jim. Crist Mortician who donated funeral services for Jane Help.
Newton, George. First husband of Elizabeth Esther Tuite, Aunt Lizzie.
Newton, George, Jr. Son of Elizabeth Esther Tuite and George Newton, who died in the Spanish American War.
Nomee, Clara White Hip. First woman to lead Crow Tribe of Montana; Bob Pelcyger represented Crows during her tenure.
Nurse.* Female springer spaniel, author's second dog in Delta, 1942-1945.
Oliphant, Dr. Manfred. OBGYN surgeon; performed micro-surgery to reverse the author's tubal ligation in July 1982.
Orson.* Name of teddy bear that became lovey of author's son; also name of twin son in fairytale, *Valentin and Orson* (written in 1489).
Pelcyger, Bob. Author's husband of forty-two years, also known as Robert Stuart Pelcyger.
Pelcyger, Eben Valentine. Son of author and Bob Pelcyger.
Pelcyger, Ellie Coben. Author's sister-in-law, wife of Joel M. Pelcyger.
Pelcyger, Eugene. Author's father-in-law, father of Bob Pelcyger.
Pelcyger, Florence. Wife of Jordan Pelcyger, daughter-in-law of author.
Pelcyger, Gale. Author's step-daughter, daughter of Bob and Deborah.
Pelcyger, Joel M. Bob's brother and author's brother-in-law.
Pelcyger, Jordan. Author's step-son, son of Bob and Deborah.
Pelcyger, Olivia. Daughter of author, adopted by Bob Pelcyger.
Pelcyger, Ruth Cantor. Author's mother-in-law.
Pepa. Nickname for Beck, Edward Robert, author's maternal grandfather.
Pierre. Carrie's first husband.
Price, Monroe and Aimee. Friends of Bob Pelcyger who inadvertently influenced him to have a late life son.
Racine. Sister of Bob Pelcyger's first wife Deborah.
Radetsky, Peter. Science writer for *Longevity Magazine* in 1993.

Rebout, Dr. Renee. Physician in Ouagadougou, Burkino Faso.
Reid-Collins, Cybele. Daughter of Judith Reid and Richard Collins.
Reid-Collins, Hilary. Daughter of Judith Reid and Richard Collins.
Reid-Collins, Oriana. Daughter of Judith Reid and Richard Collins.
Reid, Judy. ONINer and wife of Richard Collins, who called author's attention to Terry Tempest Williams' book *Refuge* in fall of 1991.
Ricks, Marc. Mormon bishop who conducted excommunication trial of author on grounds of her apostasy.
Roberts, Walter Orr. Founder of the National Center for Atmospheric Research; participant in Thuja clinical trial.
Robinson, Dr. Bill. Oncologist, melanoma specialist, at CU Medical School.
Roger. Katherine Help's fiancé.
Romary, Dr. Molly. OBGYN surgeon who performed complete hysterectomy of author in May 1994.
Rosen, Dr. John Nathaniel. Author of "The Perverse Mother" whose false theories about the cause of schizophrenia haunted the author from 1959 until 1982, born Abraham Nathan Rosen. He finally admitted he had never been trained as a psychiatrist.
Scout.* Shi Tzu dog from Patty's second marriage.
Segal, Francesca. Laura Charlotte Sternberg Smokler's mother.
Senesh, Lawrence. CU Professor of Economics, neighbor of author's father, who, along with his wife, Dorothy, raised the first litters of Pulik born in America. He loved Marley.
Sheila. Hospice nurse who conducted Ruth Pelcyger's intake interview.
Sabar, Siham. Osama bin Laden's fourth wife.
S., Dr. Oncologist who treated Jane Help, 1988-1992.
Smith, Joseph. Founding Prophet of Mormonism.
Smokler, Charlotte Sternberg. Author's neighbor and friend in Boulder.
Smoot, Nanie. Author's maternal grandmother, also known as Vermilla Smoot Beck.
Snowball.* Author's white female cat, Logan, 1948-50.
Spettigue, Beulah. Nanie Smoot's niece who died of Spanish Influenza.
Spettigue, Genevieve Smoot. Mother of Beulah; sister of Nanie (Vermilla) Smoot Beck.
Sternberg, Rubin. Father of Charlotte Sternberg.
Stevens, Peggy. Best friend of author's mother in Salt Lake City.

Stevenson, Dr. Ian. Canadian-born psychiatrist and professor at the University of Virginia, School of Medicine, who studied and wrote about reincarnation.

Stewart, Leonore Thurston. Neighbor and friend of author's father. Wife of Omar Stewart, CU Professor of Anthropology.

Stull, Dean P. Founder, Hauser Chemicals, thc first U.S. manufacturer of Thuja and Taxol.

T., Dr. Denver Oncologist who treated author and Jane Help.

Tate. Author's first born granddaughter.

Taylor, Vicky. Friend of William A. Marshall.

Teda. Author's lovey, a handmade teddy bear made by Nanie.

Tempest, Callie. Niece of Terry Tempest Williams, who coincidentally attended Boulder celebration of Frank V. Lieberman's eightieth birthday in August 1991 with Marshall Family. She was best friend of Trina DeHaan.

Thomas, Clarence. Justice of the U.S. Supreme Court, appointed in 1991.

Thomas, Nonny. EBHS classmate and best friend of author beyond.

Torkelson, Mamie." Mother's helper, who cared for Beck siblings during epidemic of Spanish Influenza in 1919.

Trump, Donald J. Elected President of United States in 2016.

Trumpeter Swan.* Saved author on shores of Yellowstone Lake.

Tuite, Amico. Twin brother of author's paternal grandmother, Anna Gertrude Tuite, who died on his thirteenth day of life.

Tuite, Anna Gertrude. Maiden name of author's paternal grandmother; twin sister of Amico Tuite.

Tuite, Bernardine. Author's paternal great grandfather; father of twins Amico and Anna Gertrude Tuite.

Tuite, Elizabeth Esther. Anna Gertrude Tuite's paternal aunt, who became her substitute mother after death of both her parents.

Tuite, Francis Torquacious. Nicknamed Torty, paternal great uncle of author, business partner of Jacob Lieberman in the T and L Coffee House.

Tuite, Frank. New Jersey police detective and cousin.

Tuite, Robert Parnell. Paternal great uncle of author, older brother of Anna Gertrude Tuite, whose wife died during the Spanish Influenza.

Turvey, Dr. Edward. Cardiologist who treated both of the author's parents.

Valentine. Given name of author's son, father, and great grandfather. Also the name of twin brother in the fairytale, *Valentin et Orson*, written in 1489.

Verhaag, Mildred. Substitute mother and manager of Bridal Salon, Brock's Department Store in Bakersfield.

Waite, Olive E. Jamaican-born health aide who cared for Ruth Pelcyger.

Wallway, Beatrice Gilfert. Author's Delta nurse and substitute mother.

Wallway, Donna. Daughter of Beatrice Gilfert and Marvin Wallway.

Wallway, Lonna. Daughter of Beatrice Gilfert and Marvin Wallway.

Wallway, Marvin. Mining engineer sent to Delta by War Department to search for uranium; shared office space with author's father and became close friend of Lieberman family.

Wallway, Mary Kay. Daughter of Beatrice Gilfert and Marvin Wallway, who was author's first friend in Delta, 1942- 1945. Given name at birth was Mary Katherine.

Wallway, Patty. Daughter of Beatrice Gilfert and Marvin Wallway.

Weinberg, Dr. Norris H. Psychologist at UCLA who provided key consultation with author in 1960 about her mother's schizophrenia.

White, Vanna. Female hostess on *Wheel of Fortune*, a television game show.

Wilkinson, Benjamin. Son of Ann Amundson and Charles Wilkinson.

Wilkinson, Charles F. ONINer and husband of Ann Amundson. Author of many books including *The Eagle Bird,* which included the chapter on Yellowstone.

Wilkinson, David. Son of Ann Amundson and Charles Wilkinson.

Wilkinson, Philip. Son of Ann Amundson and Charles Wilkinson.

Williams, Terry Tempest. Author of *Refuge* and many other books, including *Leap*, about Hieronymus Bosch's triptych the *Garden of Earthly Delights.*

Winnicott, Dr. Donald Woods. British pediatrician turned psychoanalyst who wrote eloquently about the role of mothers in the life of their children. His writing was the inspiration for *Optimal Distance.*

Work Mistress. Symbolic persona for Bob Pelcyger's professional work demands.

Yellen, Maxwell. Ruth Pelcyger's Polish-born boyfriend in Hollywood, Florida, 1983-1993.

Young, Elaine. Substitute mother for author in Logan, 1948-49.

Zegans, Lauretta. Childhood friend of Ruth Cantor Pelcyger, who lived near her in South Florida.

Zidel, Harris. Friend of Nonny Thomas in Garberville, California, who contacted author while being treated by Dr. Gonzalez for pancreatic cancer.

Zidel, Sophie. Daughter of Harris Zidel who was born the same year and month as son of the author.

Credits for Photographs and Illustrations

All photographs and illustrations in Part Two of *Optimal Distance, A Divided Life,* are from the collections of the Beck-Lieberman family, the Cantor-Pelcyger family, or are in the public domain, unless otherwise noted below.

Page

19 Gale's gift box to the author, July 1983
Photograph by Scott S. Miller, 2017

23 Manx the cat from *Colorado Daily*, November 24, 1983
Reprinted with the permission of the *Colorado Daily*.

44 Dr. J. Einhorn bringing the first Pulik to America in 1939.
Family collection of Lawrence and Dorothy Senesh, 1995.

51 Addie Help's Self-Portrait in May 1989
Reprinted with the permission of Addie Help.

305 Rubin Sternberg and Laura Francesca Segal and their two children, Emanuel and Charlotte, Lviv, Poland, January 1939
Reprinted with the permission of Charlotte Sternberg Smokler"

318 Bear cradling swan beneath the Angel Tree
Illustration by Emma Komlos-Hrobsky, June 2017

Endnotes

1. Dr. John Nathaniel Rosen was born "Abraham Nathan Rosen" on November 22, 1902 in Brooklyn to Samuel Rosen and Estella (Mollie) Aripatch, changed to Oriwitz. In the 1924 yearbook for Syracuse University, his name is "A. Nathaniel Rosen," a junior Liberal Arts major. It appears that Dr. Rosen's only medical degree was in pathology from James Madison University in 1927, although no definitive record for this degree has been found. Under oath in a legal deposition he stated: "My knowledge of psychoanalysis was limited to what I had learned in the first months of my personal analysis and from my reading on the subject." He told others that he underwent personal analysis with Dr. Ferdinand Nunberg. No records for an individual by this name have been found, but a Dr. Herman Nunberg was president of the New York Psychoanalytic Society in the early 1950s. Although Rosen claimed to have been on the faculty of Albert Einstein College of Medicine, no documentation has been found of that employment or many of his other educational and professional claims. Since he was still using Abraham as his given name in the 1940 U.S. Federal census, it is possible that some academic records were not found if the search was for John N., J. N. Rosen, or John Rosen.

Regardless, Rosen managed to persuade both the Rockefeller Foundation and the Doris Duke Foundation to contribute funds to establish the "Institute of Direct Analysis" at Temple University School of Medicine. Among his patients were Anne Lindbergh and Roger Annenberg, the son of Walter and Lee Annenberg, who committed suicide while under the care of Dr. Rosen. He surrendered his license to practice medicine in April 1983 after being charged with 102 violations of the Pennsylvania Medical Practices Act. He died in Boca Raton, Florida on March 30, 1993.

2. According to Jewish law, at age thirteen Jewish boys become morally and ethically accountable for their actions. *Bar mitzvah* means "son of commandment." For Orthodox and Conservative Jews a girl becomes a *bat mitzvah ("bat*" means daughter) at the age of twelve. For Reform Jews the age is thirteen.

3. "Morphic resonance" is a term coined by Rupert Sheldrake, a controversial English researcher and author in the field of parapsychology, who worked as a biochemist and cell biologist at Cambridge University from 1967 to 1973. While "morphic resonance" is not accepted by the scientific community as a real phenomenon, Sheldrake was an early believer of epigenetic-like phenomena, i.e. "memory is inherent in nature" and that systems like termite colonies, plants, and mammals inherit a collective memory from all previous things of their kind. He theorized that this phenomena was responsible for the "telepathy-type interconnections between organisms, ideas that involve precognition and telepathy. Sheldrake also points out the similarities between morphic resonance and Carl Jung's collective unconscious with regard to collective memories being shared across individuals.

4. D.W. Winnicott calls the teddy bear or other such object that becomes special to an infant or young child, the first "not-me" possession. One of Winnicott's most important insights was that such objects act as a preparation for separating and weaning. See "Transitional Objects and Transitional Phenomena," *Playing and Reality,* London: Psychology Press, 1971, pp 111-123.

5. According to Arthur Dickson, the first prose text of *Valentin et Orson* was printed at Lyons by Jacques Maillet in 1489.

6. "Valentine, thou art the son of Alexander,
Emperor of Greece. Thy noble mother,
Bellisant, is sister to King Pepin.
Her husband cruelly banished her and in
The wood of Orleans, she soon gave birth
To thee, and to one other on this earth.
The fellow, Orson, that thou leadest here
Is thy twin brother. He was taken by a bear
That raised and nourished him. Unknowingly,
Thy uncle Pepin rescued thee that day
And nobly brought thee up. In much distress
Thy mother roamed. The giant Ferragus
Protected her for twenty years. Now, though shall wed
The Lady Clerymond. And if thou cut the thread,

The bit of flesh that binds thy brother's tongue,

Then he shall speak as plain as anyone."

7. Suzanne Arguello, of Animal Behavior Associates in Fort Collins, Colorado, was Manx's feline therapist.

8. Kurt Chandler, "Shrink Soothes Troubled Beasts," *Colorado Daily, Weekend,* November 25-26, 1983, p. 3.

9. Among the founding couples of the Friends' Condominium were several with synchronistic connections to the author's family. Omar and Lenore Thurston Stewart were connected in two ways. The author's father and uncle attended high school with Lenore in Salt Lake City and Professor Omar Stewart testified as an expert anthropologist for Bob Pelcyger in his representation of the Pyramid Lake Paiute Tribe. Professors Kenneth and Elise Boulding lived in the condominium across the hall from the author's father. Kenneth had been a mentor to the author. Further, she had also done consulting work for Elise while she was president of the Women's International League of Peace and Freedom. When Eben Pelcyger was fourteen, he met Kenneth's and Elise's granddaughter, Meredith Graham. Eben and Meredith had a four year long-distance romance. Another founding couple was Professor Lawrence and Dorothy Senesh, who raised the first Pulik in America.

10. Tavistock Institute, founded in London in 1947, became the focal point in Britain for psychoanalysis and the psychodynamic theories of Sigmund Freud, Melanie Klein, Carl Gustav Jung, and R.D. Laing.

11. Lieberman, Joan C., "When Cutbacks Expose Hidden Problems: Case Study of a Legal Services Program," Larry Hirschhorn and Associates, *Cutting Back*, San Francisco: Jossey-Bass Publishers, (1983) 337-365.

12. At that time, the treatment protocols for the author's tumor type were minimal. According to the National Cancer Institute (NCI) guidelines, if the surgeon got a clean margin around the boundary of the tumor, radiation therapy was optional. Having stopped lactation to reduce her circulating hormones, the author decided, with the support of her husband Bob, her surgeon, gynecologist, and a consulting oncologist, to skip the radiation, choosing instead "to be followed closely" to avoid the daily trip to Denver since Boulder was then without a radiation center.

13. *The Dead Bird* (Addison-Wesley, 1938) is Margaret Wise Brown's story of a group of children who find a dead bird in the woods and decide to give it a proper, if morose

and melancholy, burial. "The children felt with their fingers for the quick beat of the bird's heart in its breast. But there was no heart beating. That was how they knew it was dead. And every day, until they forgot, they went and sang to their little dead bird and put fresh flowers on this grave."

14. La Leche League is a nonprofit organization that promotes breastfeeding through mother-to-mother support, encouragement, information and education. Founded in 1956 in Illinois, the League now has groups in sixty-eight countries.

15. Both Nonny and the author's sister-in-law Ellie Coben Pelcyger developed breast cancer. The wife of a NARF attorney had died from the disease and three of the author's former clients were in treatment and another had died. Jane, Margaret, and the author joined a breast cancer support group that included over twenty women, including Nuhiela Audeh, who became a close friend.

16. The A.K. Rice Institute is a community of teachers, students and practitioners of a discipline known as "Group Relations," a powerful and unique methodology for understanding how our unconscious thoughts and feelings significantly impact our lives when we are in groups—from family to workplace to nation.

17. The marriage was between Bruce R. Greene and Susan Rosseter Hart. Bruce was one of the original NARF attorneys. Susan, the author's secretary-bookkeeper at Head Start, followed her to NARF, remaining there for twenty years.

18. Wilber, Ken, *Grace and Grit* (Boston: Shambhala Publications, 1991), p 323.

19. Friedrich Nietzsche explained sublimation in *Human, All Too Human*. "There is, strictly speaking, neither unselfish conduct, nor a wholly disinterested point of view. Both are simply sublimations in which the basic element seems almost evaporated and betrays its presence only to the keenest observation." Nietzsche, Human, p. 38145.

20. *Bambi, a Life in the Woods*, written by Felix Salten was first published in Austria by Ullstein Verlag in 1923. An English translation by Whittaker Chambers was published in North America by Simon & Schuster in 1928. (Chambers, an American writer and editor, renounced Communism after his early years as a Communist Party member and Soviet spy.)

21. Ginny Jordan wrote a book about her experiences with breast cancer and other illnesses in *Clear Cut—One Woman's Journey of Life in the Body*. Brooklyn, Lantern, 2012.

22. Sadly, the author's paternal aunt, Alice Liebermann Marshall, was diagnosed with lobular carcinoma within months of the metastasis found in the author's liver. Alice, a

life-long resident of Utah, was then finishing chemotherapy, even though she had undergone bi-lateral mastectomies in an unsuccessful effort to avoid chemotherapy. She and her husband, Arthur Marshall, generously provided the author with five thousand dollars to help cover a portion of Dr. Gonzalez's fees. As far as the author's insurance company was concerned, she was still a Hospice patient.

23. The patient was Myra Greene Karson. She died in a New Rochelle Hospital on February 2, 1991 at age forty-six. Myra was the wife of Barry M. Karson and the mother of two sons.

24. "The Word of Wisdom" refers to a revelation supposedly received by Joseph Smith at Kirkland, Ohio on February 27, 1833 and subsequently published as Section 89 of the *Doctrine and Covenants*. In practice, during the author's childhood it meant abstaining from smoking, all drinks containing alcohol or caffeine, coupled with only sparse consumption of meat.

25. Friends' School now serves over two hundred students in three facilities, the preschool, the primary school, and a new middle school that opened in the fall of 2016. It has been selected as one of the "50 Best Private Elementary Schools in the U.S." by GreatSchools.org.

26. Paul Helmke, a Republican, was elected to three terms as mayor of Fort Wayne from 1988 to 2000. Several years after his service as mayor of Fort Wayne, he became president of the Brady Campaign to Prevent Gun Violence. He currently is Professor of Practice and Director of the Civic Leaders Center at Indiana University in Bloomington.

27. The Search Conference convened by The Greater Fort Wayne Consensus Committee took place in the Walb Memorial Ballroom on the Fort Wayne campus of Indiana University on June 27, 1991.

28. The Prudential premiums were then almost a quarter of the family income of the author's family, outrageously high, in part because many breast cancer patients were seeking expensive bone marrow transplants. Even though the author sent Prudential a notarized affidavit promising never to seek a bone marrow transplant, Prudential was still withholding approval for anything except Hospice care, while simultaneously raising premiums.

29. Williams, Terry Tempest, *Refuge—An Unnatural History of Family and Place*, New York: Pantheon Books, 1991.

30. *An Oral History of the Tuite, Lieberman, and Marshall Families – The memories of Frank V. Lieberman, Alice Lieberman Marshall and Arthur William Marshall,* transcripts of interviews with Joan C. Lieberman in Boulder, Colorado, February 1992. (Unpublished manuscript).

31. In July 2000, the Colorado State Board of Psychologist Examiners suspended Glenn Clifford's license for three years and ordered him to complete graduate-level courses on client-therapist boundaries and other issues. Also, the patient with whom he had an affair filed a civil suit against him.

32. Dr. David Luce is a Harvard trained internist who practices medicine in Boulder. He agreed to follow the author locally when she began the Gonzalez program.

33. As recreated in "The Bear Goes to Topaz." in Part I, page 14.

34. The author's diary shows the date of her telephone conversation with Bea Wallway was March 23, 1992. Bea drafted her letter the next day. It was delivered to the author's home on March 26, 1992.

35. Wallway twins death certificates

DEPARTMENT OF COMMERCE
BUREAU OF THE CENSUS
450 STATE OF UTAH
CERTIFICATE OF DEATH
State File No.
Registrar's No. 18

1. PLACE OF DEATH:
(a) County Millard
(b) City or town Delta
(c) Name of hospital or institution: (If outside city or town limits name precinct)
Delta Hospital
(If not in hospital or institution give street number or location)
(d) Length of stay: In hospital or institution (Specify whether
In this community years, months or days)

2. USUAL RESIDENCE OF DECEASED:
(a) State Utah (b) County Millard
(c) City or town Delta (If outside city or town limits write RURAL)
(d) Street No. (If rural give location)
(e) If foreign born, how long in U.S.A. years.

3 (a) FULL NAME (Baby) Wallway Infant Wallway
3. (b) If veteran, name war
3. (c) Social Security No.
4. Sex Female
5. Color or race
6. (a) Single, widowed, married or divorced S
6. (b) Name of husband or wife
6. (c) Age of husband or wife if alive years.
7. Birth date of deceased Feb. 24 1945 (Month) (Day) (Year)
8. AGE Years Months Days 1 If less than one day 12 hr. 5 min.
9. Birthplace Delta, Utah (City, town, or county) (State or foreign country)
10. Usual occupation
11. Industry or business
FATHER 12. Name Marvin Joe Wallway
13. Birthplace Sioux City, Iowa (City, town or county) (State or foreign country)
MOTHER 14. Maiden name Beatrice Eleanora Gilfert
15. Birthplace Emerson, Nebraska (City, town or county) (State or foreign country)
16. (a) Informant's own signature Beatrice E. Wallway
(b) Address Delta, Utah
17. (a) Burial (Burial, cremation, or removal) (b) Date thereof 2 27 45 (Month) (Day) (Year)
(c) Place: burial or cremation Delta
18. (a) Mortuary
(b) Signature of funeral director
(c) Address Delta (d) License No. 140
(e) Was body embalmed? No (f) Embalmer's License No.
19. (a) Mar. 9 1945 (Date received local registrar) (b) (Registrar's signature)

MEDICAL CERTIFICATION
20. DATE OF DEATH (Month, day, and year) Feb. 25, 19 45
21. I HEREBY CERTIFY, That I attended deceased from Feb 24, 1945, to Feb 25, 1945;
I last saw her alive on Feb 25, 1945;
death occurred on the date stated above, at 12:00 pm.
Immediate cause of death
Undeveloped heart and lungs
Due to Premature birth
Due to
Other conditions (Include pregnancy within 3 months of death)
Major findings:
Of operations
Of autopsy
Duration
PHYSICIAN Underline the cause to which death should be charged statistically.
22. If death was due to external causes, fill in the following:
(a) Accident, suicide, or homicide (specify)
(b) Date of occurrence
(c) Where did injury occur? (City or town) (County) (State)
(d) Did injury occur in or about home, on farm, in industrial place, in public place? (Specify type of place) (e) While at work?
(f) Means of injury
23. Signature (M.D. or other)
Address

DEPARTMENT OF COMMERCE
BUREAU OF THE CENSUS
450 STATE OF UTAH
CERTIFICATE OF DEATH
State File No. 20
Registrar's No. 19

1. PLACE OF DEATH:
(a) County Millard
(b) City or town Delta
(c) Name of hospital or institution: Delta Hospital
(d) Length of stay: In hospital or institution
In this community

2. USUAL RESIDENCE OF DECEASED:
(a) State Utah (b) County Millard
(c) City or town Delta
(d) Street No.
(e) If foreign born, how long in U.S.A. years

3. (a) FULL NAME (Baby) Wall Way Infant Wallway
3. (b) If veteran, name war 3. (c) Social Security No.
4. Sex Male 5. Color or race W 6. (a) Single, widowed, married or divorced S
6. (b) Name of husband or wife
6. (c) Age of husband or wife if alive years
7. Birth date of deceased Feb. 24, 1945
8. AGE Years Months Days If less than one day 1 hr. 20 min.
9. Birthplace Delta Utah
10. Usual occupation
11. Industry or business
FATHER 12. Name Marvin Joe Wallway
13. Birthplace Sioux City, Iowa
MOTHER 14. Maiden name Beatrice Eleanora Gilfert
15. Birthplace Emerson Nebraska
16. (a) Informant's own signature Beatrice E. Wallway
(b) Address Delta, Utah
17. (a) Burial (b) Date thereof 2 27 45
(c) Place: burial or cremation Delta, Utah
18. (a) Mortuary L.N. Nickle + Son
(b) Signature of funeral director Albert F. Nickle
(c) Address Delta, Utah (d) License No. 160
(e) Was body embalmed? No (f) Embalmer's License No.
19. (a) Mar 9 1945 (b) Leola A. Roberts

MEDICAL CERTIFICATION
20. DATE OF DEATH (Month, day, and year) Feb. 24 1945
21. I HEREBY CERTIFY, That I attended deceased from Feb. 24 1945, to 19
I last saw him alive on Feb. 24, 1945
death occurred on the date stated above, at 3:15 p.m.
Immediate cause of death Undeveloped heart and lungs [illegible]
Due to Premature birth
Due to
Other conditions (Include pregnancy within 3 months of death)
Major findings: Of operations
Of autopsy
22. If death was due to external causes, fill in the following:
(a) Accident, suicide, or homicide (specify)
(b) Date of occurrence
(c) Where did injury occur? (City or town) (County) (State)
(d) Did injury occur in or about home, on farm, in industrial place, in public place? (Specify type of place) (e) While at work?
(f) Means of injury
23. Signature M. E. Reid (M.D. or other)

36. Beatrice Gilfert Wallway was born June 12, 1913 in Emerson, Nebraska to John Gilfert and Katherine Levis. She married Marvin Wallway on June 6, 1934 in Union, South Dakota. Bea and Marvin moved to Los Angeles in 1936, where Marvin attended the University of Southern California and Bea worked at a hospital. Their first daughter, Mary Kay (Katherine), was born there on February 22, 1942, before they moved to Delta, Utah in July 1942. After the loss of their twins in February 1945 and the end of World War II, Bea and Marvin returned to California, where their daughter Donna was born in December 1945. They eventually settled in Casper, Wyoming, where they had two additional daughters, Patty and Lonna.

In August 1996, following Marvin's death, the author drove to Casper and had a poignant reunion with Bea. At that time Mary Kay, married with three children and living in Philadelphia, was in remission from breast cancer. Sadly, Mary Kay died on January 3, 1999 at age fifty six. After the death of Mary Kay, the author and Bea spoke and corresponded several times between 1996 and 2003. Beatrice Gilfert Wallway died in her sleep on August 17, 2015 in Sioux Falls, South Dakota at the age of one hundred two. Bea is survived by her three daughters, eight grandchildren, and four great grandchildren.

37. Dr. Hans C. Moolenburgh began practicing medicine in 1953 in the city of Haarlem in the Netherlands. Although he was a general practitioner, he developed an interest in clinical ecology and subsequently began treating cancer patients. The author of several books, he helped lead the effort to end fluoridation in the Netherlands.

38. Originally published by Uitgeverij Ankh-Hermes, Deventer, Netherlands with the title *Een engel opje pad.* Moolenburgh, Dr. H. C., *Meetings with Angels—A hundred and one real-life encounters*, Saffron Walden, Essex, England, C.W. Daniel Company, Ltd., 1992.

39. In general, Frank V. Lieberman followed the guidelines of Dr. Nathan Pritikin, which were based on a low-fat diet and moderate aerobic exercise. In many aspects, Frank's diet was more conservative than recommended by Dr. Pritikin.

40. Two days before the wedding of Bob and the author, Frank V. Lieberman had been moving loads of gravel in a wheelbarrow to create a path for the backyard ceremony. He had not complained of pain, but afterwards he vomited and passed out in downtown Boulder. He refused to get into the ambulance when it arrived, telling the attendants, "I'll be fine; my daughter is getting married and I don't want to create a disturbance." Two weeks later, Frank went to visit his sister Alice in Salt Lake City, whose brother-in-law, Dr. John Marshall, insisted Frank be evaluated. After an angiography was performed on a Friday afternoon at St. Mark's Hospital, Frank was told to put his affairs in order and return Sunday for open-heart surgery on Monday. The author flew to Salt Lake on Monday morning. Her plane was delayed because of thunder storms so she arrived after Frank had been taken to surgery. She was surprised to find not only her Aunt Alice in the surgical waiting room, but her mother's two older sisters. Frank underwent a quadruple by-pass and came out of surgery converted to rigorous dietary restraint.

41. For further information see Amin-Ud-Din, M; Salam, A; Rafiq, MA; Khaliq, I; Ansar, M; Ahmad, W (Mar-Apr 2007). "Aposthia: a birth defect or normal quantitative recessive human genetic trait?" *Eastern Mediterranean Health Journal.* 13 (2): 280-6. PMID 17684849

42. *Sophie's Choice*, a novel written by William Styron was published by Random House in 1979. A film of the same name was produced in 1982 in which Meryl Streep played the lead role of Sophie, a Polish survivor of the German Nazi concentration camps. At gunpoint, Sophie is forced to make a horrendous choice between the survival of her son or her daughter.

43. On June 6, 1994, Dr. Gonzalez presented the author's case history and all of her medical records as one of three of his best cases at a conference sponsored by the National Institute of Health's (NIH) Office of Alternative Medicine (OAM) in Bethesda, Maryland. A report of his presentation was published in the *NIH OAM Newsletter*, AM, Vol. 1, Number 6, July 1994. The published report contains several minor factual errors; whether they are those of Dr. Gonzalez or the editor is unknown.

> *"J., a 50-year-old businesswoman, underwent left breast lumpectomy for carcinoma in November 1986. J. Did well until July 1989, when her physician detected a mass in her right breast. A lumpectomy documented poorly differentiated adenocarcinoma. An abdominal ultrasound revealed a density in the right lobe of the liver; a needle biopsy confirmed carcinoma.*
>
> *J. Began CAF chemotherapy, but in November 1989, after completing 13 cycles, she refused further treatment. At that point there had been no improvement in her liver lesions. For several months she did nothing. She then learned of my work and, in April 1990, she began the program. After two years on her protocol, she felt so well that, without my knowledge, she discontinued the protocol. In July 1991 she suffered a grand mal seizure; a CAT scan revealed two brain lesions.*
>
> *J. Immediately resumed her full program, and showed rapid improvement in all symptoms. CAT scans of both the head and the abdomen on April 17, 1992, were completely normal and she remains well."*

44. *Schmatte* is a Yiddish word often used to refer to discounted or low quality goods. It comes from the Polish word "*szmata*" meaning "rag."

45. Patients with tumors containing receptors for estrogen and/or progesterone can be treated with hormone therapy, including drugs like Megace and Tamoxifin that block female hormones. Tumors lacking hormone receptors are considered to be more aggressive.

46. Colorado Uninsurable Health Insurance Program (CUHIP) was a state-sponsored high-risk health insurance plan that began operating in 1991 to provide health insurance to individuals unable to purchase health insurance at affordable rates from commercial insurance companies due to pre-existing conditions. The program ended after the Affordable Care Act became law.

47. Myriad Genetics, Inc., an American molecular diagnostic company based in Salt Lake City, Utah, was the first to offer a test for the BRCA gene.

48. When the author had an MRI to measure the impact of "Anti-Her 2" as per the requirements of the clinical trial, the radiologist thought he saw metastases in four of the author's cervical vertebrae, but a more cautious neurosurgeon was less certain. He proposed exploratory surgery and proved his diagnostic skills. There was no sign of bone metastases; instead there was a loose bone chip from her third cervical vertebrae, plus three others so riddled with osteoarthritis that he replaced them with ones from a cadaver. The surgeon then used two titanium plates to stabilize her neck. The aftermath of the hay truck accident seemed never-ending, but the absence of bone metastases was a great relief and good news for future stockholders of Herceptin.

49. *The Substitute Wife* was a television movie first shown on NBC in 1994; directed by Peter Werner; the script was written by Stan Daniels.

50. Soon after arriving in Santa Cruz, Margaret tagged along with her mother and step-father on a European tour focused on religious studies. On the tour, Margaret began a love affair with an American man in the tour group unusually susceptible to Margaret's golden aura. The initial weeks of their love affair acted like a powerful stimulant on Margaret's immune system; there may have even been a few moments of *optimal distance*. She returned to Boulder radiant, with plans to see her new lover again in California, full of eager impatience for their romantic reunion. It turned out that her new lover was a student of Mary Magdalene and may have been sick from his own perversion. Whatever it was, within hours of reconnecting in California, he began telling Margaret of his many encounters with prostitutes. Margaret flew into a hormone-free rage. He had re-wounded her, using Glenn Clifford's contaminated sword.

51. Ian Stevenson, M.D., *Twenty Cases Suggestive of Reincarnation*, Charlottesville, Virginia: University Press of Virginia, 1974.

52. On July 30, 2015, Osama's step-mother, Rajaa Hashim, his half-sister, Sana bin Laden, and her husband, Zuhair Hashim, died when their executive jet crashed in Southern England.

53. During the period when the author was caring for Ruth in Florida, Ruth's younger brother Laurence Cantor, the keeper of the Cantor family tree, called frequently to check in with Ruth. As Ruth's health declined and her need for sleep increased, the author often took his calls and they began conversing about family history. After Ruth died, Laurence

asked the author if she would assume responsibility for carrying on his work after his death. She agreed, while pointing out that Laurence's life expectancy was longer than hers, but promising to find a successor if she died after him. (Laurence died May 22, 1998).

54. Aunt Mary died July 1, 2006 at age ninety-four from injuries sustained in an accidental fall down the basement stairs of her Salt Lake City home.

55. The author subsequently learned that wildfires put fireman at high risk for a stroke due to a combination of the stress, exertion, and dehydration. Further, paramedic firefighters are specifically trained to identify and treat the symptoms of a stroke.

56. Joseph Smith claimed that he used a pair of stones, fastened to a breastplate, to assist him in the translation of the golden plates. He subsequently referred to the stones as "Urim" and "Thummin"

57. There has been considerable controversy about the baptism of Jews by Mormons, but the practice of baptizing Holocaust victims is particularly objectionable. At one point both Anne Frank and Hitler were listed in the LDS International Genealogical Index (IGI). Attempts have been made by Jewish leaders and prominent Jewish genealogists to persuade LDS officials to limit names submitted for baptism to those tied to a current church member. Signed agreements have been broken.

58. The Mutual Improvement Association (MIA) is a component of the Mormon Church that provides educational and social opportunities for young men and women. The purpose is to ensure each young person is worthy of making and keeping sacred Mormon covenants. Groups of young people, divided by age and gender meet weekly on Tuesdays for recreational and faith-promoting meetings, as well as to undertake various self-improvement projects.

59. Valentine Liebermann died in Belleville at age seventy six. No death record for Tecla has been found. She was listed in the 1920 City Directory for Orange, New Jersey as "widow of Valentine" where she was a boarder at 2 Rutgers Street. It is possible she remarried, but as of 2016, no record has been found for any subsequent marriage.

60. Orphan trains were the creation of The Children's Aid Society. Beginning in the 1850s and continuing to 1929, the Children's Aid Society sent many abandoned, poor, and/or homeless children from New York City to new homes in rural America on orphan trains. The leader of the Children's Aid Society, Charles Loring Brace, hoped that farmers would not only take these children into their homes, but treat them as their own. Many were paraded like slaves before prospective parents on train platforms.

61. Elizabeth Esther Tuite had many secrets. She was married twice; first to George B. Newton, with whom she had a son. In the 1880 U.S. Federal Census, she was living in Virginia City, Nevada hotel, working as a "hair dresser." She listed her marital status as "widow" and reported that she had given birth to one child. She was a guest in the same boarding hotel as Edmund Joseph Harris, single, whose occupation was "baker." The couple married in Virginia City in 1881. Her son, George Newton, died in 1897 in Havana, Cuba from injuries suffered in the Spanish American War. Newspaper reports from that era stated that her first husband was still living and was a resident of Caracus, Venezula.

62. Text of the letter sent by Francois du Verne, a French citizen, to Anna Tuite Liebermann in Ogden, Utah on August 4, 1919:

> *"Dear Mrs. Lieberman,*
>
> *"You'll be surprised to hear from me for you do not know who I am. I'll explain it in two words, this afternoon, while walking in my park, I saw a bunch of Americans. A few minutes afterwards, I had the chance of meeting your husband with whom I had a good talk. As I guess that you'll be glad to have news, I write to you. First of all, I must say that he seems to be getting on very nicely. He spoke of his home tenderly and thinks of it all the time—perhaps even does he find that France is too far from the United States. I must say too, that I had not to spend a very long time with him, before appreciating him very much. I find him to be so upright and gentle that I hope we shall write and meet again. Unhappily, he only had a long stop at the station that is quite close to my home and the train has gone up to a further camp and it will not be so easy to meet. However, be sure that I am very glad to have made the acquaintance of your husband.*
>
> *"Yours sincerely, Fr. Francois du Verne*
> *Chateau du Veuillin par le Guetin, Cher (France)."*

63. Research indicated that Donald Liebermann was referring to Sarah Ann Butler, the wife of Veronica Moran's younger brother, Joseph William Moran.

64. "Our parents make separate but equal genetic contributions to who we are. They also make separate but unequal epigenetic contributions. For some genes it makes a difference whether you inherit them from your father or your mother. These genes are

epigenetically activated if they come by way of your father (and vice versa). Other epigenetic states, some environmentally induced, can be transmitted from grandparents to grandchild." Francis, Richard C. *Epigenetics–How Environment Shapes Our Genes.* New York: W.W. Norton, 2011, xiii.

65. In a 2013 study researchers trained mice using foot shocks to fear an odor that resembled cherry blossoms. Later, the extent to which the offspring of those mice startled when exposed to the same smell was tested. The offspring had not even been conceived when their fathers underwent the shock/scent experience, and had never smelled the odor before the experiment. Their reaction to the cherry blossom scent was about two hundred percent stronger than that of a control group. The scientists then looked at a gene called "M71" that governs the functioning of an odor receptor in the nose that responds specifically to the cherry blossom smell. The gene, inherited through the sperm of trained mice, had undergone no change in its DNA encoding, but it did carry epigenetic markers that caused it to be "expressed more" in descendants. This in turn caused a physical change in the brains of the trained mice, their sons and grandsons, who all had a larger glomerulus – a section in the olfactory unit of the brain. Dias, Brian and Ressler, Kerry, "Parental olfactory experience influences behavior and neural structure in subsequent generations," *Nature Neuroscience*, Vol. 17, Number 1, January 2014, page 89 95.

66. Kumsta, R. Hummel, E., Chen, F.S., & Heinrichs, M. (2013) "Epigenetic regulation of the oxytocin receptor gene: implications for behavioral neuroscience." *Frontiers in Neuroscience*, 7,83.

http://doi.org/10.3389/fnins.2013.00083

67. For example, "*The Hongerwinter*" (hunger winter) of 1944 caused the death from starvation of eighteen thousand citizens of Holland during World War II after Germany prevented the shipment of food supplies. Children born during the *Hongerwinter* were small and underweight at birth, compared to siblings born before and after the *Hongerwinter.* A similar pattern was found in Nigeria for the children of the 1968-1970 Biafra famine.

In the case of the *Hongerwinter* children, they showed a below-average methylation of the insulin-like growth factor II gene (IGF2), which codes for a growth hormone critical to gestation. Later studies in this cohort found increased methylation of five other genes, among them genes associated with cholesterol transport and aging. Subsequently a careful review of the meticulous Dutch medical records showed those exposed to the Dutch famine prenatally had a significant increase in the risk for schizophrenia. Heijmans BT, Tobi EW, Stein AD, et al. "Persistent epigenetic differences associated with prenatal exposure to famine in humans," *Proceedings of the National Academy of Sciences of the United States of America.* 2008, 105(44):17046-17049. PMC. 12 Feb. 2016

68. Jha HC, Mehta D, Lu J, El-Naccache D, Shukla SK, Kovacsics C, Kolson D., Robertson ES, 2016. Gamma herpesvirus infection of human neuronal cells. *mBio* 696):e01844-15.doi:10.1128/mBio.01844-15

69. Takako, Tabata, et al. "Zika Virus Targets Different Primary Human Placental Cells, Suggesting Two Routes for Vertical Transmission" *Cell Host & Microbe,* Volume 20, Issue 2, August 2016, 155-166.

About The Author

In many ways, the roots of *OPTIMAL DISTANCE, A Divided Life*, extend back to June 17, 1947, the day after my fifth birthday. As I lay on my stomach on a surplus Army blanket in Logan, Utah, I used a crayon to illustrate my feelings in a pink diary with a praying child printed on the cover. Unable to spell more than a few words, I tried to draw how I felt. The diary had arrived that day in the mail, a birthday present sent by my Aunt Mary from Salt Lake City. It had a pseudo lock and key, but I soon lost the tiny aluminum key and had to break the lock. When the pink diary had no more empty pages, I began using a school notebook, a habit that persisted for the next seven decades. Writing has always kept me alive.

I was born in Utah in 1942. My father was a scientist and atheist, distantly related to Simon Bamberger, the first and only Jewish governor of Utah. My mother, a descendant of prominent Mormon pioneers, lost her familial faith during the Depression. Tragically, she developed paranoid schizophrenia shortly after my birth and, from then until her death, her mind was under the control of invisible demons. As an only child in Logan, Utah, I took refuge in the interstices of the Mormon Church trying to keep a safe distance from my mother's unpredictable and murderous impulses because I knew she would never follow me there. I often felt like a small wild animal desperately hiding from danger among a large herd of domineering dairy cows.

I was fourteen when I left Utah and Mormonism behind after my father was transferred to Bozeman, Montana. A year later he was sent to Bakersfield, California where I finished high school. Following a year of study at the University of California, I traveled in Europe and worked as a medical volunteer in Africa. When I returned to Berkeley, I gave birth to my daughter, pushing her stroller through demonstrations for Free Speech and against the War in Vietnam. In 1966, I left Berkeley to finish my thesis on leadership in a rent-free house in Northern Idaho. Two years later, a job offer to become the director of the county-wide Head Start program brought me to Boulder, Colorado, where

I still reside. In 1971, I undertook an assignment for the Ford Foundation to help develop the Native American Rights Fund, where I met and fell in love with Bob Pelcyger. We were married in 1975, and for the next four decades, I worked as a management consultant serving clients who were lawyers, doctors, and women in leadership positions.

In 1999, as a finalist for the Bakeless Literary Prize, I was invited to attend the Bread Loaf Writers' Conference. My publishing consult was with Carol Houck Smith of W.W. Norton, who encouraged me to expand my workshop submission into an autobiography. Then fifty-seven, I had been "living" with metastatic cancer for a decade, and didn't believe I had enough time left on my life clock to undertake such an effort. Nonetheless, buttressed by Carol Houck Smith's endorsement, I began searching through my diary entries for clues and areas of research. Progress slowed after I suffered a near fatal stroke while flying to Utah in 2007. Sadly, Carol Houck Smith died the day after Thanksgiving in 2008, while improbably, I lived on.

In August 2014, our thirty-ninth anniversary, Bob and I made a bucket list for our marriage before our approaching deaths. The only item on Bob's list was the completion of *OPTIMAL DISTANCE, A Divided Life*. Now seventy-five and dependent on a life-sustaining cocktail of medications, I have finally fulfilled my beloved husband's request. Credit should go to Bob for his endless persistence and to Aunt Mary for the gift of a pink diary.

CPSIA information can be obtained
at www.ICGtesting.com
Printed in the USA
LVOW06s1602271117
557737LV00024B/495/P